Blind Vision

Blind Vision

One Family's Story of Treasure, and the Greed of Outside Forces

Mark Chibbaro

Blind Vision
One Family's Story of Treasure, and the Greed of Outside Forces

v3.0

Cover courtesy of artist Lauri Blank

Outskirts Press, Inc.
http://www.outskirtspress.com

ISBN: 978-1-4787-5596-8

PRINTED IN THE UNITED STATES OF AMERICA

Preface

Pick an object in a room. Close your eyes and think about it for a moment, then hold it in your hand with your eyes closed. Take a moment to get a sense of its texture, composition, and even its color(s). If you are fortunate to have someone in the room that will allow you to caress their face in this way then you will likely "see" them in a whole new way. I was fortunate to be the son of a blind man and a woman who had a wonderful heart; she never saw any disability in his blindness. Both had the ability to make everyone around them feel they were the most important people on earth. Their union provided a foundation from which they could serve others while building wealth for those they loved, to show the world that what may be perceived as a disability may truly be an advantage if understood and managed to one's fullest. Wealth for them meant that the bills were paid, your health was good, and that your loved ones were close.

This is a theme you will experience along a wild ride in the text.

The book you are about to read in based on the life of my father. He was born in 1924 in New Brunswick, New Jersey, and died in a small town just a few miles away, Milltown, in 1989. There are many things about this book that are fact and some that are utter fiction. He built a substantial business with his brother and was married to a woman named Rose who was a CPA. They both had parents who immigrated to the United States to build better lives. My grandparents loved our country and the opportunity it afforded those who wanted to work hard. In fact, my parents were forbidden to speak the native language of their parents out of respect for the people of the country that they now lived in. The values of hard work, perseverance, and love of God, family, and country were certainly bestowed by my grandparents to my parents and by extension to my brother and me.

On his deathbed my father had only one wish: he wanted his family to stick together no matter what was going on in their lives. As a family unit we never felt that we had a family member who was disabled in any way; rather, we felt special to have such an independent and accomplished man in our presence. He was so loved that when he died the funeral home had a line outside that stretched around the block for two days. In fact, many people came back several times to pay their respects.

Many key elements of what you will read are true. My father was conservative and my mother was somewhat liberal. They respected each other's views but while they differed in this way it never got in the way of the family. They independently grew their own businesses. It is true that my mother convinced my father to get involved with treasure hunting in the Florida Keys. To be a son of a conservative businessman and a mom that was a liberal treasure hunter was quite the dynamic.

Some of the things that occur in the book are fiction. I will let you in on one of them: my dad did not own or fly a plane, but we would sit outside for hours on the weekends and I would explain them to him as they flew overhead. As far as aviation, though, the Hindenburg experience did happen for him. This is but one example of the fact versus fiction of the manuscript. You will notice very quickly that certain names have been changed.

My father was the best role model a son could have hoped and prayed for. His sense of humor was keen.

His command of his profession and the compassion he had for others shined in his daily life. He rarely raised his voice as he used the gift of words and actions in a way that I could spend a lifetime trying to describe. He never complained about anything and when he walked into a room it lit up. The space around him literally changed. The positive energy of this wonderful man touched thousands and I hope you come away from this book feeling his love and spirit; I hope you feel uplifted.

The following, which was titled, "Our Brother," was written by his siblings as his eulogy:

"To a brother who has been an inspiration to us with his faith in God, love of his wife and sons, his unending selflessness, his deep concern for all of us and in his suffering.

The benefits we received from assisting him in any way will be with us forever. He appreciated us at all times. 'Thank you, thank you,' he responded to us. He showed never-ending patience with family, friends, and coworkers, yet his concern was never to impose upon us in any way. There was never a complaint.

Now, for his brilliant mind and memory: only the Lord knows. It is UNEXPLAINABLE. Monsignor Harding once said: 'You go to the Pope for religion and Matty Chibbaro for insurance.'

Our pain and grief at this time can only be relieved by knowing that his sight is restored in his eternal life and happiness.

Now, it is our turn to say 'Thank you, thank you,' Matty, for showing us the way; you are our leader and our idol."

Table of Contents

Chapter 1:
DECEPTION

May, 1985: a black Suburban pulls up to the Park Avenue home of Mathew Messina. In the back is the blind executive, Matt, who is adjusting his tie and reaching to grasp the bouquet of exquisite pink peonies he had his florist fly in from Columbia . Matt has striking good looks. His salt and pepper hair flows back in perfect waves. As the sleek limo approaches the door of his building it slows to carefully align with door, guarded by the building's business-like doorman, so that Matt can exit the car without any assistance of seeing-eye dog or cane. Matt would not hear of such aids. New York City doormen are a blend of trusted confidante, security guard, local statesmen, and close friend. They all see no evil, speak no evil, and hear no evil. As Matt exits the car and walks toward the doorman, Shamus knows the ritual of his walk, a walk that would remind one of the precision of the orchestrated march at the Tomb of the Unknown Soldier in Arlington National Cemetery.

As Matt approaches the door he extends his hand at the same moment Shamus does; a simple, "Good afternoon," is exchanged by each man. There seems to be something different about the sensation of their handshake this day; it causes Matt to pause a moment before entering the opulent foyer of the stately uptown residence. He leans over to smell the peonies and a smile comes over his face as he enters the elevator to the penthouse. Shamus watches the door close behind Matt then reaches for the page box and hits the button for Matt's apartment in a clear effort to alert someone upstairs; little does Shamus know that the buzzer in the apartment is broken.

After Matt keys in the proper code his hand extends upward and he opens and closes his fist, thinking again about the less-than-earnest handshake he had exchanged with Shamus. As the door to the residence opens Matt takes another moment to enjoy the scent of the flowers he is about to present to his wife. He hears something in a distant room and he calls her name but there is no reply. For a moment he pauses and then walks to the kitchen and calls her name again, but still there is no reply. He begins to grow concerned and heads for the bedroom while repeatedly calling her name. He raises his hand, just as he did in the elevator while recalling the doubtful handshake with Shamus, and then places it on the handle and slowly opens the door.

"Dawn, are you there? Dawn, honey, I can smell your perfume; I know you are there. Are you playing one of our fun little games?"

Just as he says this he stops in his tracks. He senses there is something wrong and walks to the edge of the bed and pauses for what seems to be an eternity. He places the roses on the nightstand and is frozen in time. He can hear the faint sounds of the breath of two people in the bed. Dawn and her lover are trying to be as still as possible but know that he knows. Matt walks around the bed and faces the 20-foot-high wall of glass that overlooks Central Park and the distant top of the George Washington Bridge. He paces back and forth and decides to go back to the bedside to gather his peonies. He turns to the bed.

"I will never think of peonies the same again."

The smell in the room was a mix of the odors of sex and the lovers' clashing perfumes. He approaches the door and without turning around bows his head to the floor; the peonies fall from his hand and he raises both hands to within one inch of his face. He is now opening and closing his fists just as he had done in the elevator. He turns his head without turning his body and says, "Dawn, do I have to fire your brother Shamus or would you prefer to?" There is a gasp from the bed and Matt turns and walks away.

The lovers are like cadavers and the chill that has come over them is mirrored by the storm clouds rolling across the sky outside. The sound of the elevator door slamming is in concert with the thunder as Matt makes his way back to the lobby. This time Matt is somewhat less certain about the direction of his walk as his agitation has caused his precision navigation to be less than perfect as he glances off a marble statue and a flower pot before confronting Shamus. At this point, Shamus sees him struggling and offers a hand and Matt declines, saying, "Haven't you already done enough damage for one day?" Matt orders Shamus to summon his driver, falls back on a couch in the vestibule and drifts into a depressed comatose state.

Chapter 2:
REFLECTION

Matt begins to reminisce about his childhood and the school for the blind that he had attended in Jersey City, New Jersey.

The building was an old, turn of the century brownstone staffed by wonderful, dedicated, loving nuns. It was a mild day in May of 1937 in his memory, on the dusty streets of Jersey City, as the black Buick Special driven by Matt's rigid Italian immigrant parents prepares to drop him off at the front of the building. They are very proud and proper people with very little to say and they are particularly proud of the family's first automobile. When they spoke the words were directed and powerful. Matt's father gives his son a small duffle bag and wishes him well while his mother embraces her precious son, saying, "Remember, Matteo, you are a Messina; be strong and respectful. As always, your father and I will be here next week to bring you home for the weekend." A nun greets Matt and walks him into the school, where there is a bit of a commotion taking place through the vestibule and halls. There is some confusion, and teachers gather in the halls. There is talk of a German airship that is about to pass over; the children are directed to go to the rooftop play area so they can hear the rumble of the motors as it passes closely overhead.

The children are escorted swiftly up the narrow staircase; they arrive just in time to experience the historic event. The image for those with sight is ominous and powerful. For those that cannot see, there is a feeling of awe and curiosity. There is not a sound on the rooftop as the children anxiously listen to the Hindenburg approach. The nuns try to describe the image to the blind students but they seem to be overwhelmed themselves. As the low-pitched hums of the motors draw closer, Matt reaches up to the sky with his hands open as if to feel the blanket of mild wake created by the massive airship. It glides by with silky grace and dignity. The swastikas emboldened on its sides shout the message of a proud Germany. The image of blind schoolchildren beneath the behemoth paint a picture of future deception yet to be learned by the world. The experience settles deep within Matt's soul.

As the airship passes the children are led back down

the stairs, but Matt manages to stay behind and listen and ponder well after the sounds of its humming motors can no longer be heard. A feeling of confusion comes over Matt. He senses something is very wrong and faces up to the sky. Just as he begins to feel the heat of the sun on his face one of the nuns shouts his name and takes him by the hand to lead him back down the stairs and into his room. Matt settles into his modest bunk and begins to study his daily braille lesson. Just a short time later he hears the cries of the nuns, who are just learning of the tragic end of the Hindenburg. Matt stands up slowly and puts both hands against the window, as if to reach out for the doomed airship. He falls back to the bed the exact same way he would fall onto the bench, years later, in the vestibule of his apartment after learning of his wife's deception. It was in this moment, as a small child, that Matt began to understand his ability to sense things in a manner that others did not.

Matt is woken by his driver, who is unsettled by the sight of Matt sitting there so disheveled. Matt's driver, Aleximov, is a massive man from Bulgaria who is also one of his bodyguards and security staff members. The two get in the car and Matt asks them to pause for a moment. He asks the driver to get Shamus to the car. Shamus is led to the car by the driver. Matt lowers the window a few inches after hearing two double taps on the window, which is their known code. Shamus is asked to leave; a replacement would be there later that day. Shamus attempts to speak more but Matt raises the window. Matt asks Aleximov to drive him to his home in the Hamptons. Matt opens a crystal flask and begins to drink several glasses of single malt scotch, after which he passes out; the drive takes hours longer than it should due to an accident on the Long Island Expressway.

While passed out in the car Matt dreams of his childhood back in New Brunswick, New Jersey.

Matt had loved sports, particularly baseball and stickball. In his dream, he is in the street playing stickball with his brother and other local boys of Italian descent in their working-class section of town. The families are very proud of their modest homes and possessions. Most had come to the United States with very little, and to have the ability to rent or own a home is taken very seriously. Matt's father, Salvatore, is returning home in the one suit he owns after a day of selling insurance door to door. Salvatore is a proud, imposing man, with striking good looks. It has been a very long day for Salvatore. He works everyday from 8 a.m. to 6 p.m., except on Sunday, which for Catholics was the Lord's day of rest and reflection.

It is the end of his workday, and he sits on the stoop of the apartment building where they live to watch the children play. Matt's brother becomes incensed over a dropped catch and begin shout in Italian. Salvatore jumps to his feet and drags Matt's brother, Isidore, into the small and very neatly kept apartment. He reprimands his son for speaking in Italian and lectures that he and his family are never to speak Italian in public, as it is disrespectful; the only place for their former language is strictly within the family's home.

Matt is still outside and begins yelling for help as he is having trouble seeing. In a matter of weeks, Matt's sight is gone forever. His parents take him to several physicians but there is nothing that can be done. They are told it is a rare form of glaucoma. Matt's mother and father both work to make ends meet so it is decided that everyone in the family will care for Matt. He is doted on by a loving and caring family of one brother and three sisters. Everyone in the family has a role in helping him.

Matt's father takes him to work once each week, and it is there that he listens to his father sell insurance to families. After each day of work, Matt's father stands at the window overlooking the street below with his arms outstretched and hands pressed against the glass, to connect with the pulse of the world around him—a trait that Matt adopts as well. Matt quickly realizes this business is something a blind person could do well and begs his father to take him more and more. Matt is a quick study and this learning forms the basis for an amazing career. Matt's father is a strong, principled man of imposing stature. They have very little and family means everything.

These values become the core of Matt's fiber. He models himself after his father. Each day his father says, "Matteo, my son, each day you must do something nice for someone and always stay away from trouble. If you sense trouble, walk away."

Matt begins to wake up just as the car turns into the long driveway of his beachfront estate in the Hamptons.

Chapter 3:
REALIZATION

As Matt begins to come to his senses the realization of the day's events begin to sink in. A few of the petals from the roses remain on the car seat next to him. The scent disgusts him to the point where he starts to punch the seat. Showing concern for his flustered state, Aleximov suggests that he should assist Matt to the residence. That offer is vehemently refused as yet another single malt is poured by Matt, but this time he is shaky and spills all over the car seat and his beautifully tailored suit. As Matt exits the car he brings the glass with him, creating a trail of spills as his shaking continues. The tremors are the manifestation of the effects of the booze and shock of the afternoon's events.

At this time, Matt's assistant of thirteen years, Damiano, is alerted to the arrival by the barks of Matt's three golden retrievers: Claire, Beatrice, and Carmella. Matt had named them after his three favorite teachers at the school for the blind. There was a fourth dog as well, a proud German Shepherd named Frankie after a childhood friend of Matt's who had died at a young age. He was always there to help Matt when he needed it.

Damiano had been dressing in the second-floor quarters of his own room and rushes to finish tying his tie. The three beautiful dogs rush for the door. As it is opened, Matt is greeted by the loving attention of his dogs. Claire has a tendency to be a bit more clingy and protective than the others and bumps into Matt, causing him to fall against the wall, where a painting of Dawn hangs. The glass of scotch is flung backward and the rather strong libation splatters over the image of her face and begins to run down her body, causing streaks of shame.

The home is a spectacular, custom-built center hall colonial with twenty thousand square feet, which had been designed by Matt's first wife, who had died of cancer. Matt had offered to move to a different vacation home or have Dawn redecorate after they were married but she refused. The only change she had made was to commission and hang a huge picture of herself in the front of the home as to announce she was the new queen of the castle. This, of course, was never too popular with anyone—especially Matt's children.

Damiano rushes down the stairs and is appalled by the sight of Matt in this state. As was the case in the car, Matt refuses any assistance and the two men make their way into the great room, which has huge floor-to-ceiling windows overlooking the Atlantic Ocean. Damiano pauses to look at the painting and his eyes show concern and anxiety. He has never seen Matt like this before. He turns to see Matt with his hands outstretched on the large windows overlooking the beautiful beach and breakers just outside. The pulse of the ocean was palpable on the glass. It amplified the beat of Matt's heart, and he was sure that the dogs sensed it as well; they whimpered and gazed at Matt with loving eyes.

Frankie passes the window and sees the anxiety in Matt's face; he makes his way into the home via the dog door in the pantry. As Frankie approaches, Matt's hands slide down the glass as his legs become week. He curls up against the widow on floor in a fetal position. Damiano reaches out to clasp Matt's left hand to help him up as the dogs sigh and whimper; Damiano guides Matt to a chair. The two men are now seated facing each other as Matt continues to open and close his fists. Damiano is frozen in his seat as Matt tells him of what just happened in the New York City home. Damiano is more than an assistant to Matt; he is a trusted advisor and close confidant. He is many years older than Matt and was a friend to Matt's father as well. He asks Matt if he had any idea who the man was. Matt says no, however, he mentions that he thought Shamus was involved in some way, noting that it seemed Shamus had forgotten that that the buzzer in the apartment was broken since his attempt to alert the lovers had gone unnoticed. In fact, as Matt thought more of the events of the day he suggested that Shamus might have turned off the security camera. As the revelation of conspiracy begins to take shape, Matt stands up, walks to the wet bar in the corner of the massive living room, and pours himself a heaping glass of single malt scotch. The bottle is positioned in a manner and has such a distinctive shape that Matt is able to discern what it is. Damiano asks Matt not to drink anymore, and as the words come out of his mouth he says, "Screw it; would you mind if I joined you?" "Damiano, you might as well. I think we are going to have a long night ahead."

The two walk outside to sit in two Adirondack chairs perched above a beautiful dune that separated the home from the beach. The dogs follow and settle around them on the ground. Matt thinks about keeping the news of this away from his two sons for the time being and asks Damiano where they might be. Damiano explains that they had called earlier and were staying at the country club for dinner after a round of golf. Matt's sons, Mark and Luke, ages eighteen and twenty, were home for summer break. Matt's daughter, Rose, who was twenty-two, was in Palermo visiting with relatives in Sicily while studying abroad. Mark was a gifted writer and musician but was pursuing a career in medicine. Luke was finishing he degree in civil engineering

"Damiano, I want you to call Jeff, my lawyer, first thing in the morning. I want him to contact my security people and find out who this guy is. At this point I only want to know his name and his background. Also, tell them what happened with Shamus, as I am certain he is in on this now in some way. Enough of this for now, I will think about the next steps once I know more about this man."

Matt continues to drink and begins to reminisce of the wonderful times he had shared with his family in the beach home.

He shares stories with Damiano as the night wears on about his dear wife, who had died of breast cancer six years prior. Of course the entire family was devastated; Rose was the one who had been Matt's emotional cornerstone throughout his wife's illness. Matt loves each of his children equally, but it is Rose who can see directly into Matt's soul.

Matt tells Damiano a story of one winter day where snow had blanketed the yard. Rose loved the snow and had asked to play outside. While she was running though the snow she noticed a flower had pierced through the top layer of snow. It was a tiny purple flower. Rose got on her knees and carefully removed the snow from the stem; she removed the flower from its cold harbor, cupped it in her little hands, and ever so carefully walked into the house where she had noticed her father sitting in his favorite reclining chair listening to classical music. She took his hand and placed the delicate little bloom in his open palm, and said, "Daddy, please smell this; I found it for you." Matt was overwhelmed with joy. His little girl had found something of beauty in the cold, dark depths of winter and wanted to share it with him. The two had connected on a spiritual level that only a father could appreciate in its entirety. At that moment Matt knew he had the basis for a speech for her future wedding day: the man lucky enough to get a girl who finds a beautiful flower in the throes of a cold, snowy, grey day and makes sure her father enjoys the delicate texture and scent is surely going to be blessed himself.

Matt is clearly more focused on surrounding himself with people who love him than wanting to speak of the affair he had walked in on earlier that day. Matt instructs Damiano to lock down the front gate after the boys return and to make no mention of what had happened in the Manhattan home until he had given more thought to it.

"Damn it, I think Rose tried to warn me about Dawn and never really spoke up," Matt says.

"Matt, everyone was more interested in seeing you happy and moving on with your life. The entire family was concerned that Dawn was nineteen years younger than you, and that she seemed to have a tendency to be self-absorbed, but we put those thoughts aside as we saw how you seemed to be so happy around her," Damiano replies.

"I am to blame for poor judgment; I cannot believe my stupidity. Dawn had portrayed herself as a devout Roman Catholic Christian, yet always had an excuse when it came to going to church services or events," Matt says. "Damiano, I should have seen this coming. I had asked Dawn on several occasions to come to Haiti to help us with our Health Foundation and she always seemed to have an excuse. Funny thing though: she always managed to go to the black tie affairs. She was more interested in getting her face in the society pages than helping people."

Matt and Damiano speak well into the night as the

sun sets in front of the home and the sea begins to grow louder and more angry in front of them. Mark and Luke return from their dinner to find the alcohol-stained painting of their stepmother and notice spilled alcohol on the floor. As they walk through the living room they notice that the door to the lanai is open and head out to the pool area, where they see their father and Damiano now sleeping in the chairs. They wake the two and ask what has happened. Matt simply says it was a hard day at work and that he chose to come out to the beach home to get away from things for a couple of days. The boys want to know more but Matt insists everything is ok. They all head into the home and the troop of faithful dogs follow. Damiano brews a pot of fresh coffee for the men and they head into the library.

The boys know something is very wrong and ask where their stepmother is. Matt replies that she is back in Manhattan as she had some events to attend. They spend the evening chatting about their favorite hobbies: soccer and cars. They each have their own favorite teams but agree that the governing body that controls soccer on a global basis has serious credibility issues. There is some idle banter on which teams are better but they are unified on the topic of cars, Maserati being the clear choice for them. The men choose to move their little gathering into the garage, where Matt walks over to his favorite vehicla. He runs both hands across the hood of his black Maserati Quattroporte. In a very quiet voice his says, "Gentleman, only an Italian can build the feel of a woman and the power of Caesar into a machine such as this." He instructs Damiano to start the car and asks his sons to close their eyes and put their hands on the hood of the car. They are now all at the hood, eyes closed, with hands gently laying on the hood. He says, "Boys, listen carefully to the beat of the motor and feel the heart of this beauty. This is the soul of the car speaking to us; whether it is living or not it has an energy and a soul."

He then runs his hands slowly down the side of the car with the side of his face, almost caressing the fenders and doors as he makes his way to the back of the car. Matt has a keen sense of humor and it often catches people off guard. This is such a moment, as he says, "Men, the allure of this machine is the sound that hums from these tail pipes. Don't confuse the sound of anything that comes out of any other tail." That comment generates laughter, like men in a locker room sharing dirty jokes. Matt knows that he has shocked the boys with the remark and apologizes for the bad analogy. He is clearly still wound tight from the events of the day and wants to internalize the matter until he decides his next step. So he says, "Well, boys, what I meant to say is that we need to listen very carefully, as sometimes a car can be like a lady: your best friend or your struggle. She will send you signs, and most of the time if you treat her with great care she will take you places you never dreamed. If we treat all things with respect, dignity, and care, everything in our lives will fall into place."

Matt goes on to say that they have been blessed with many material things but nothing is more important than family and that although their mother had departed this life years ago her soul lives on in all of them. Matt's father had instilled the principle of family in the most Italian tradition: stick together and look after each other. He always instructed his boys that no matter what happened in life you must stay connected on every level. God and family are your only foundation; everything else is built on shifting sand. The men finish their coffee in the garage, seated at a small workbench. Matt tells the men he is exhausted from a long day. He takes a few steps toward the door and says, "I want you all to do something going forward. Please don't let a day go by without doing something nice for some one every day, no matter how small the gesture, and don't let a day pass without telling someone you love them. Let me be the one to start."

He tells them he loves them, then Matt heads back into the house. The dogs follow Matt, as they know this is the time of day where they get an evening treat. Matt heads for a small room off the kitchen and it is there that each dog is fed a treat and they then retire to their respective pooch beds. He embraces each of the animals and then heads up the impressive staircase to his bedroom.

Matt closes the door behind him and goes to a window that faces the ocean. He opens it and sits in a chair next to the window for a few minutes as the sea breeze blankets his face and the sounds of the waves comfort him. He is starting to get tired and sits on the bed and reaches for a picture of his family on the nightstand. It is a wonderful picture of his first wife and family on the beach with matching clothes: khakis with white button-down shirts and bare feet. He clutches it and sits on the bed and pulls the picture up to his chest where he holds it to his heart with a death grip and cries himself to sleep.

Matt dreams of his children when they were very young: it is Christmas morning, and the children are opening his presents with him holding his wife's hand. They enjoy every second of the presents being unwrapped. His daughter turns to him and walks over to sit on his lap. Matt places his hands gently on her face and his fingers feel the contours of her features. He is overcome with the joy of the moment and says, "You are as beautiful as your mother." The sound of wrapping paper being torn away from boxes grows silent and the only sound in the room is the crackling of the fireplace. The boys look on at the tender moment and sit next to him on the couch. One by one, Matt sculpts a picture of each child in his hands as he gently touches each of their faces, ending with his wife. He tells the family that this is the best present he could hope for: to have his family and to be able to create a sense of their unique beauty in his head, which forms a blind vision in his mind. He says he is blessed with being able to see things in a way sighted people are not and this is a gift from God. He gets up off the couch and walks to a vase filled with Christmas flowers. He runs his fingers slowly up the vase and then caresses the stems while leaning forward to inhale the scent. A smile of joy fills his face with a glow of holiday comfort.

Morning sun breaks though the bedroom window. The heat wakes Matt, and as he rolls to the side the picture of his family that he had fallen asleep clutching crashed to the floor. The frame and glass are shattered and scattered. He sits up and sweeps his feet back and forth to clear the mess away and a shard of glass penetrates his foot. He begins to shout expletives, not only from the pain of the glass but rather the recollection of the prior day's events. He places his hand on the intercom to summon Damiano to help him and then sits very still as the morning sun bathes his face. The large windows frame a spectacular view of the rising sun and the ocean below. The glass splinter in his foot seems forgotten as Matt sits still and allows the glow to warm his face. Matt reaches for the remote that controls the sound system in the room; the sweet, soft sounds of Mozart put Matt in a trance. The broken shards of glass glisten and sparkle almost in concert with the sympathetic mood in the room. Damiano knocks at the door but there is no reply from Matt. He calls Matt's name and begins to shout and knock louder. Damiano can hear the sound of the music but nothing else. The door is locked and he decides to kick it in, the commotion of which wakes up the boys down the hall. Damiano rushes in and sees Matt seated upright, facing the window. He gingerly walks around the mess on the floor and speaks Matt's name in a soft, reassuring voice. At this moment the boys arrive at the bedroom door and witness the confusing but tender moment. Damiano asks the boys to fetch a first aid kit as well as some things to clean the floor. They dutifully rush to gather the items. They rush down the stairs, and when they are far enough away he asks Matt when he going to tell the boys. Matt explains that he has given some thought to this and will do so over breakfast on the family yacht that afternoon. The boys return and clean the mess on the floor and Damiano tends to the wound on Matt's foot. Luke begins to ask what is going on and Damiano holds up his hand as to motion the boys to back off with the questions.

Matt shares his plan for the day on the yacht. Fishing is good this time of year and the weather is great. He sends the boys off to gather their gear for the day on the sea. Damiano is sent to help the boys get ready. After they all have left the room, Matt approaches the window with a noticeable limp from his wound. This time he only places one hand on the window with the other behind his back. He speaks out loud to himself in a very stern monotone: "Be patient and be at peace as a plan will come to me." Damiano pages the room via the intercom to announce the Suburban is prepared and ready, directly out the front door. Matt replies that he will change and be down shortly and tells Damiano that Aleximov can have the day off if he likes. The boys pass Matt's room and offer their assistance but Matt tells them to head out to the car. He changes into his boating attire and slips into his pocket an antique Italian pocketknife that his father left him when he passed. He walks down the stairs but takes pause next to the damaged painting of his cheating wife. He reaches into his pocket to fetch the small blade and takes a few deliberate cuts at the painting in *x* patterns. He folds the knife in a very deliberate fashion and places it in his pants pocket. He pauses briefly to get his composure; with this the front door swings open and Damiano looks at the painting and then to Matt and asks if he if he should dispose of the painting before the boys see it. Matt says, "No let's tend to that later; by the time we get back the boys will likely know the whole story." The men leave the residence and join the boys who are waiting in the car.

Chapter 4:
Shooting the Moon

Once they are on the way to the marina, Mark strikes up a conversation about their last trip on the boat when he had caught the most fish. The men get into a fun, spirited spat over Mark's recollection, stating each had caught the most. Matt stops the conversation, reaches for his pocketknife, and while repeatedly opening and closing the blade he says, "You know, men, it's not the amount of fish you catch, it's the amount you set free alive to be caught again." There is silence in the car as the men ponder the statement and stare at the enigmatic grin on Matt's face. Matt reminds Damiano to call the groundskeeper to walk and feed the dogs in their absence. It is a short ride to the marina and it is a beautiful morning. As the men pull down the gravel-paved entry, Matt rolls the window down and inhales the familiar scent of salt and fish in the air. The men are greeted at the dockside parking spot marked with the name of their sixty-foot Viking fishing yacht named: "Forget a Boat It." Their captain, Billy, has been their trusted helmsman for over ten years and he greets them in his usual cheerful manner. He speaks with deep Irish brogue and tells the men conditions are near perfect for a great day of boating and fishing. Mark is still curious and asks if anyone has heard from their stepmother. Matt says he is sure she will check in soon. The engines are active, humming in the background, and they board rapidly to head out of the inlet into the open sea. Matt tells Billy to rev each motor slowly up and down at the dock before they leave as he thought he had heard something wrong in the stern side motor but was not sure. Billy is certain there is nothing wrong as he tends to the yacht with the greatest of care. He also knows Matt's senses are better than any diagnostic tool he has when it comes to things like this so he smiles at Matt and the men and executes on Matt's request. Nothing is heard but as Billy begins to glance back at the gauges on the helm he notices that the port side engine is showing that a warning light is lit, but all the gauges seem correct. He utters some Irish words to himself and Matt hears the babble. Matt asks Billy, "What is the concern?" Billy says, "It all looks good, but I will go down below to have a look at the engines and support equipment." Damiano decides that that this may be a good opportunity to run home and get Frankie. He asks Matt if it is ok, and he agrees that is a great idea as Frankie loves being out on the water. With that, Damiano leaves to get the beloved rover. Matt takes this moment to climb up the very steep ladder, up to the crow's nest, which is very high above the yacht and used for observation; there is also a second set of controls there. Once up in in the perch he leans back to feel the rising sun on his face. After a moment he picks up the intercom to page Billy in the engine room. Matt instructs him to come up to the lower helm if he doesn't find anything wrong and to run the yacht close to shore on just the one engine that has no warnings showing. Billy agrees this will be a safe way to run the yacht and returns to the main deck and begins preparations to leave the dock. Damiano returns quickly and as soon as the back gate of the Suburban is open Frankie jumps out in excitement and runs to the boat and the boys help him aboard. All are now on the yacht and the lines are set free.

Billy idles the majestic yacht out of her slip. Matt remains in the crow's nest with his hands resting on the controls, which mimic Billy's inputs below, as the two systems are linked to command the yacht in the same manner. The sleek vessel glides through the marina into the open water of the bay. Matt orders Billy to expedite their passage to the open sea due to their lost time dealing with the engine. In just a few minutes they are in the Atlantic Ocean. Billy was correct when he had stated the conditions were perfect. Matt asks all below if there are any boats in view and all confirm none are in sight. Matt requests a heading that will take them south, past the house, but that he will control the speed of the boat while it is on directional auto pilot; he also asks Billy to fire up the second engine. The men look at each other and in their delay Matt tells Billy, "Do it now; I have owned this beauty far too long to have never done this." Billy glances at the others, who affirm the decision with simple nods, so he plugs in the course and heading which then locks into autopilot. He announces over the intercom, "Matt, you have her now." Matt whispers to himself, "Indeed, I do, I really have her now." Matt grasps the throttles to both engines and with deliberate yet even temperament advances the engine RPMs just as a pilot would do to roll an airliner down the runway toward lift off. Matt enjoys the sensation of wind and salt spray on his face as the speed increases. The bow begins to rise and soon they hit maximum speed. The men below keep a keen look out in all directions for any obstruction whatsoever, whether it be debris or watercraft. Billy advises Matt that they are fifteen minutes from their destination given the current speed. Matt acknowledges Billy and leans back in his captain's chair with one hand on the helm and the other on the binnacles. The sensation of the sun on his face with the hum of the engines coming through to his fingers on the chrome of the throttles along with the warm wind against his face sends him into a blissful trance.

He begins to daydream of moments on the beach with his children and first wife and the happy times they had there. His wife lies on the ground with her head in his lap as he strokes her hair while the children play with Frisbees and beach balls. The moment is utterly euphoric. Her eyes close and Matt moves his fingers to her face to sculpt the view of her beauty in his mind. He tells her that every man should do this, as a woman's beauty cannot truly be appreciated with the eyes alone. Every muscle tells a story, as they form the living reflection of the soul though the most gentle of motions to the most intense contractions. He asks her to close her eyes and feel his face and experience this with him. As this is happening the children take notice and slow their play to witness this tender moment. They gather around the blanket and both parents embrace the children with a family group hug that seems to last an eternity.

Matt is awakened by the shouts of Billy, who

quickly takes control of the yacht, as they had almost come into the path of another fast boat from their port side. A collision is avoided and Billy slows the vessel down as the family home comes into view off the stern. Matt removes his hands from the controls and is noticeably shaken by the near miss and from being wrenched from such a sweet dream. He orders Billy to drop anchor in front of the home just a few hundred yards from the shoreline, where the waves are breaking against the sandy Hampton beach.

Matt makes his way back down the steep ladder and asks all the men to gather together for a toast. "Men," he states, "here we are on one of the finest fishing yachts ever made in front of our beautiful beachfront estate. None of this means anything without family. You must stick together no matter what happens in life. There is no greater bond than family, men, and I would like to ask each of you what you would do if someone inside the family did harm to one another." The boys look at each other with concern and Mark replies first: "Dad, I suppose it depends what it was." Matt jumps in with a stern voice: "Let me be clear: nothing should be tolerated, not even a bad word about each other. If you don't have a good word about someone—especially family—don't say it at all." Luke gestures to Mark, who is clearly schooled, and says, "We understand, dad. Is something bothering you? You seem to not be yourself." With a nervous laugh Matt says, "Hey, sorry for the heavy stuff, men, and so, for today, lets raise our glasses to family and to a wonderful day on the water." They finish their drinks there and ponder what Matt has just said. Billy is asked to raise the anchor and take them offshore to do some deep sea fishing.

As the boat begins its journey Mark goes below deck to get some privacy. Down in the main salon he communicates with his sister via special equipment designed for use on the boat to connect with phones on land. "Hey, sis, how is your trip coming?" Rose tells him that he must make the trip over; she has been very happy with her experience in Sicily.

"Family here means so much, and our relatives do so many things together. Family here feels like it did before mom died," she said. "There is a sense of true love for each other, and everyone does things each day to support each other, such as chores, meals, church, and lots of family discussions."

She urges Mark and Luke to come and join her for the remainder of the trip.

"I would like to, but dad seems to be upset about something," Mark says. "Has he called you, Rose?"

"I haven't spoken to dad in several days, and in

fact, he seemed to be in very high spirits the last time I spoke to him."

Mark goes on to explain that dad seems to be acting out of character and that he was acting a bit weird.

"One minute he is as happy and philosophical as he always is, but the next minute he has a tinge of anger," Mark says. "Dad is always so even and mellow; there is something not right."

"That's strange," Rose says. "Everything seemed just fine when I spoke to him several days ago. In fact, dad had mentioned that he was planning to take more time off from work and spend time with Dawn and all of us. He went on to say that he wanted to speak with us as it related to perpetuation plans for the business before he spoke to the lawyers. You know dad; always planning. I bet he planned for us well before we were born. We are so lucky to have a father like him."

"Well, sis, it sounds like things are all good, but I had better get back to the guys." "Ok, Mark but please think about visiting me before I leave here. This is more than a trip—it's a pilgrimage, and the feeling of togetherness is overwhelming."

As they are talking, the winds start to pick up a bit and swells start to smack the sides of the yacht. A few things fall off shelves and make quite a racket.

"What is all that noise?" Rose asks.

"We are out on the boat right now and are about to head out beyond the range where our communications equipment may not be able to connect through to a land line; can you please call dad tomorrow and then call me back? Sis, you know you have a way with dad and you may pick up on something."

Rose promises to call her father and get back to Mark.

"Ok, sis, I had better go back up to the main deck now and help the guys get the fishing gear ready."

Mark pauses after he hangs up to think for a moment to himself about the call before returning topside. Billy is busy handling the boat and Matt is now perched in the teak fishing chair. The chair is a beautiful, hand-crafted seat mounted near the stern of the boat and is used to position the fisherman to be able to handle large catches. It is a throne of sorts. The teak is harvested from exotic islands and is adorned with shining chrome mounts, which place the rods in just the correct angles. It has a shining steel footrest that is used to brace the fisherman once the prey is on the line. At a cost of thirty-five thousand dollars, this is not an ordinary chair. Matt says, "Gents, this is a place of truth. This is not a chair from which to simply hook and reel a fish but rather a place of honor. Anyone can catch a fish, gents. It's all about what you do during the fight and how you conduct yourself once your prey has surrendered. You know our rule: no fish dies intentionally. Everything we catch must be returned to the sea to live another day."

Mark and Luke are busy getting all the gear ready. There is much to do. The gear is removed carefully from hand-crafted lockers and mounted with great care. Each component has a specific mounting place, almost as if to be preserved in a museum-like fashion. Billy announces they should be at their favorite location in about fifteen minutes and is on the radio listening to the chatter of other mariners in the area. The banter coming over the radio is spirited with shouts of many fish sightings. Billy laughs and reminds the men that most of that is typically a ploy to lead others away from where the real action is. Billy watches the depth gauges and fish finder screens carefully as they near the coordinates that were previously recorded in their logbooks. Schools of fish are seen about fifty feet beneath the boat and they become more prevalent as they approach their favorite location. Years prior, Billy and his friend had sunk an old boat, which created a perfect haven for bait fish and predators to hold in the area. Only a handful of men know where this is, and once they are well out of sight of others they begin to position the boat to drift over the man-made mini reef.

The lines are readied and in the water quickly as Billy set the yacht to drift over the reef. Matt smiles; he is quick to remind the boys that he will catch more than anyone. Luke jumps right in and says, "Dad, with all due respect, I think this is my day; I am feeling it." Mark seizes the moment and says, "The only thing you are feeling is seasick; I will catch the best fish." The men have a short laugh and things begin to get business-like very quickly. They all are familiar with the routine. Matt places his right index finger on the line as to feel any minute pulse of a nibble. The manner in which Matt handles the rod, reel, and line is like no other. One could say it is like the way a physician takes the pulse of someone; he gently rests his index finger and middle finger on the line. Billy says, "Look at him, he is quite the practitioner." In just a matter of moments, Matt feels the signal from the depths; something is nibbling at his bait. The sensations coming through the line are faint but deliberate as they pulse from the water below. Matt reports that this "patient" is very much alive. Matt raises and lowers the tip of his fishing rod a few times and with a sudden and violent movement he yanks the pole up and back and yells, "Fish on, boys." Matt pushes his back into the fighting and braces his feet against the footrest; he gets ready for what is to become a protracted battle between man and beast. The boys surrender their rods to their holders and stand on each side of their father to cheer him on. The men glance at each other and smile as they revel in seeing the joy in their father's euphoria. The fight seems to go on for nearly an hour. Billy continues to maneuver the boat into position to give Matt the optimal direction to reel, as the fish doesn't want to give up too easily. It is vital that the line remains directly behind the boat so as not to get snarled in the propellers or get under the boat's hull and snap the line. Matt's smile is larger than ever and the struggle becomes harder as the minutes tick by. Luke is concerned as Matt is sweating and beginning to show signs of fatigue and thus offers his father assistance. He tells the boys that it doesn't count if he hands the rod over. The rule is that the person who hooks the fish is not allowed help until it gets alongside or in the boat to be unhooked and released. Matt senses the fish is beginning to give up. He says, "Men, don't be fooled by a fish that seems like it is ready to give up. They are a lot like people; they can crafty and deceitful. One must stay vigilant and directed, as the moment you lose focus is precisely the time in the fight when things can go sideways on you."

As those words come out of Matt's mouth the fish

got a second wind and took off, pulling more and more line from the reel.

"You see what I mean, men? The fish was resting and planning its next move, but I was ready and had my drag and tension set on the reel just right to be ready for such a challenge. This is a feisty one, men, this one. When we get up close, I plan to have some words for it." Matt lets out a little chuckle and continues to reel away. The fish breaks water and the surge is felt through the line.

"Stay ready, Matt," Billy yells, "It is losing and may be its last chance to snap your line."

"Not to worry, Billy, I have her in control much better than she realizes. Yes, I do I have her now; I have that witch now."

The boys look at each other and their father without saying anything as his joy turns to anger.

"Come on, try to get away from me now."

Damiano motions to the boys to remain quiet.

"That's it Matt, you are in control," Billy says.

"Oh, yeah, I am going to teach this one a lesson," Matt says.

All are shocked by the anger and Matt's words but

let him continue the drama that is playing out. The seas continue to get rougher and Luke says, "Billy, I thought you said conditions were great today."

"All is well; these are large swells being brought in by a distant storm," Matt says. "I could care less what storm is near or far; I am going to close this deal and bring this one to the boat."

"Yes, dad, of course, we cannot wait to see you win this one," Mark says.

Matt says to ready the gaff; it may be needed as he senses the fish is beginning to give up and that is surrendering to him.

"Boys, get ready," shouts Damiano.

With a few last tugs and winds of the reel the fish goes weak and is swimming in the direction of Matt's pull on the line.

"Ok, boys, here she comes."

The men are leaning over the stern of the boat and shout that this is the largest dolphin (the fish, not to be confused with the mammal). This was an eighty-pound Mahi and the largest they had ever seen. Matt grins and says, "Men, this is our moment, not only mine. I need a few words with her before we send her on her way."

The men are overwhelmed by the enormity of the

moment and the size of the fish. Luke says, "Dad, this is the most spectacular fish I have ever seen, and I have never seen a fish with such great colors."

Matt says, "That is nature's little trick; it's like make-up. You see son, just like the women in the world, fish have the same traits: even they must be attractive in nature to attract a mate. If you were to leave that fish on the deck for a short while those colors would turn in a hurry."

Matt sits down on the stern and braces himself on the railing and says, "Honey, I hope you had a great time today. I had a better time, but your day is about to get better. I am going to let you go, but be aware I am going to come back another day and I will be ready for you. You had better get stronger and better as I will get better and stronger as well; I am done with you for now." He instructs Billy and Damiano to do what they have to do to release her unharmed. The men are shocked by Matt's demonstration and work the fish loose. Luke says, "Dad, I don't think I am feeling very well; would you mind if we head back?"

"Sure, son, lets go below."

"You gents go relax. I will take care of things out here," Billy says.

"I will put all the gear away while Billy captains; you men head down below and enjoy," Damiano says.

The father and sons sit down on large, comfortable chairs placed around a hand-carved table. Luke asks, "Dad, would you like something?"

Matt replies, "Sure, lets enjoy this moment; surprise me and pour for us.

Luke looks for the best scotch in the cabinet and says, "Ah, yes we have it." It was an unlabeled bottle of single malt which was privately sent to Matt by a family-owned distillery in Scotland. He carefully pours the scotch into three crystal glasses that bear their family initial *M.* Matt senses some tension and says, "You guys know better; I heard the squirming in the chairs. Do you have something to ask me?"

Luke replies, "Well, yes, dad. You seem to be angry; is something wrong or do you have something on your mind that you want to share with us?"

Matt puts his feet up on the table and takes a rather large gulp of the potent scotch. He keeps the glass in his left hand and begins to clench his right fist repeatedly.

"Boys, there are going to be some changes in my life. Dawn and I will be splitting up."

Matt stands to the gasps of the others and walks to the starboard side porthole and with drink still in hand he says, "Men, somewhere out there in the water is a cunning and skilled fish that decided to take my bait." His voice becomes angrier with each word. "We fought and fought and when the moment of truth came we set her free. In my case it is time to do the same."

"Dad, what happened? We never saw you two fight."

"Boys, this is a story of lies and deceit yet to be told as I don't know all the details as of yet. You can be sure of one thing, we will get to the bottom of this. After your mother died, I was resolved to live my life alone and enjoy my children, family members, close friends, and—with God's blessing—many grandchildren. I intend to do exactly that. I don't want any pity or concern about me. I will go about my business and take care of what matters most now and that is to take care of the people who really matter in our lives.

"Are you going to tell us what happened, or how you found out about this deception?" Mark asks.

"Son, there is no point in telling you what she did. Remember our golden rule: never say anything bad about anyone no matter what. That kind of thing eats away at your soul and only gets mistranslated if others repeat the message. So, that said, I will not discuss what happened. For now, all you need to know is that it involved cheating and deception, and I am disappointed in myself for not picking up any signs sooner. As for your sister, I do not want her to hear any of this while she is away on her trip. She will be worried about me and return and I do not want that to happen."

"Dad, I have an idea," Mark says. "Why don't we have the jet prepared right away and head over to join her? It would be great if we got away from all this right now and spent time with her and our cousins. We could surprise her and make this a very special time."

Matt takes another big slug of scotch and slowly turns away from the widow. With a large smile he says, "Yes, but under one condition: your sister is to know nothing is wrong. I will think of something to say as to why Dawn is not there. I just want her to enjoy her journey as much as possible. I feel you boys understand where I am coming from on this, correct?"

Matt and Luke reply in unison, "Yes, of course." Matt asks Damiano to call ahead and have the plane prepared. Matt is exhausted and tells the boy he is going to the master stateroom to take a nap. The men exchange hugs and Matt makes his way to the bed and falls asleep in a matter of a few minutes.

Chapter 5:
ENCOUNTER

Matt begins to dream about a summer's day in 1944. He is reading a braille book in his small, Spartan bedroom in the working class row house street in New Brunswick, New Jersey. There is the screech of brakes from a car in the street below. He hears the shouts of "Hey, Matty" outside his window. His closest friends are in the Black Ford Coupe that just made the noisy arrival. Inside the car are "Spider" and "Schank," two young, spirited, first-generation Italians. "Spider" was Matt's best friend; his real name was Frankie. As for "Schank," no one knew him by any other name.

"Hey, Matty, we have pop's car; let's go for a ride down to the Jersey Shore."

Matt's mother overhears this plan being hatched and she marches out the door and down the stairs in her red, sauce-stained apron with a rolling pin in her clenched right hand, smacking it in her open hand like a baseball bat.

"You boys better take care of my Matteo," she demands in her Sicilian accent. "If not, I will take care of you both with my favorite little kitchen tool."

Matt kisses his mother and assures her everything will be ok. With that the men jump into the slick new ride and head down the dusty street on their way to a day of adventure.

It's a hot day in Jersey and the boys start to hatch a plan. Schank is a young man that God did not bless with looks or social graces and Spider is a lean, good-looking young man with perfect grooming who always has a plan. Spider says, "Boys, today we head directly to the boardwalk and see what the stock of ladies is like." The drive takes them through the farmlands of Eastern Jersey. They pull off the road and drive the car behind some trees and an empty barn to enjoy a few beers that Spider had hidden in the trunk and Schank says, "Let the party begin." They have a conversation about what types of ladies they prefer. All Matt wants is a redhead. Schank laughs and says, "What does that matter to you if you can't see her?" Matt remembers his sighted days and recalls his first infatuation with a girl in the neighborhood. "Schank, you know it really doesn't matter, but it's important to me so keep your radar up for a redhead for me." Spider asks each of them what their favorite thing is, other than hair, about a woman. Schank replies, "Well, I don't know, I never been with one, and I guess I would have to say boobs." Matt says he loves the overall shape of a woman and is intrigued by every inch. Spider says, "You both are nuts; it's the butt that matters because that is the business end and most women sit on the brain." The boys have a few more laughs and get back into the car with their early afternoon buzz. They pull up to Point Pleasant Beach, where the streets are crowded with happy beachgoers, many of which are soldiers and sailors on leave from their duties of World War II. They park the car and head up to the boardwalk where the sounds of games of chance ring out and the smells of sausages and peppers on grills fill the air, along with hints of popcorn and cotton candy.

Matt's senses are on overload and he and the boys are in high gear, hunting for women. Before long, they hook up with three gals. They spotted them playing the game where you shoot water from a gun to try to pop a balloon. Spider notices one redhead and says, "Hey, red, I bet my blind buddy here can outshoot you." She replies, "I bet he can outshoot you all day long, big mouth." Matt immediately falls for her and her attitude; she is instantly enamored with Matt. Recognizing his blindness, she reaches to take his right hand with her left and raises it to cover it with both of her hands. The warmth of her heart and strength of her conviction is felt in her grasp. Matt stares directly in her eyes and she his. She tilts her head to one side and smiles. The two may as well be alone on a deserted island. The scent of her perfume, albeit not nearly as strong as the smells of cooking popcorn and sausage sandwiches, is all Matt can smell. The caress of her silky, smooth hands and her scent have put Matt into a trance.

Spider jumps in to ruin the moment: "Enough of this, let's have that shoot-out! She says, "Of course, Mr. Mouth, but not before we are properly introduced. My name is Rose and these are my friends Linda and Barbara. I propose that we team up in pairs."

"I am Matt, and this is Schank and Spider."

"Wow, at least my guy has a name," Rose says.

She leads Matt to the water gun and places his right hand on the trigger and his left under the barrel. She says, "Honey, we have this one in the bag." She walks behind him and places him in a hug as to guide him in the proper direction. The others follow in kind but do not hug the other two men. They simply place their hands on their shoulders. Schank throws a few coins on the vendor's shelf and with the loud ring of a bell the game begins. Spider is off to a lead. Matt is in third, but with some adjustments of her hips Rose guides Matt to the target and he soon takes the lead and wins the race with the pop of his balloon. Rose jumps up and down and turns Matt back to face her and says, "I knew you were a winner the moment I saw you." Matt is choked up and says, "Honey, I have never won a thing in my life but feel that the two of us could be a heck of a team."

Barbara asks how long the men are going to be there that day as they were doing an Andrews Sisters show that night. Matt is delighted,

"You mean you gals are going or are you the act?"

"Honey, we are the singers tonight," Linda says, "so why don't you come and be our guests?" Spider

"Oh no, my father is going to kill me if I don't get his car back by sundown," Spider says.

"Yeah, and even worse, my mother my chase you with her rolling pin!" Matt laughs.

"Spider, live up to your name and and let's stay; this is going to be great," Schank says.

"Well, ok, it's a plan; meet us at the South Pier

Casino entrance at 7 p.m." Rose says, giving Matt a kiss on the cheek; the others exchange a simple handshake. Matt is glued in place; as the girls walk away, Rose turns around to look.

"Guys, did she turn around or keep walking?" Matt asks.

"She turned around and looked so long she almost ran into someone," Schank says.

"Gents, let's find a place to pick up some nice shirts," Matt says. "That casino place sounds kinda nice and we should show some respect.

"Oh jeez, lover boy here is losing his mind already. Me and Schank are not doing that," Spider says

"Well, the heck with you two slugs; take me to a place where I can get something nice."

Matt is the only one who buys a nice, new button-down, resort-style shirt; the others remain in what they had worn that morning. The three make their way to the event and are greeted at the entrance by the three women who are now in their military garb, just like the Andrews Sister had been doing. Schank is overwhelmed.

"You ladies sure clean up nice," he says.

"Excuse him, I think he meant you are all beautiful," Matt says.

Rose notices Matt's shirt and says, "My guy looks great!" She locks arms with Matt and leads them to front-row seats that had been saved. "You boys relax and enjoy the show; we will see you later."

Spider whispers in Matt's ears that their parents are going to whip them good. Matt doesn't reply; he just smiles as the music begins and the curtain opens to the sound of "Boggy Woogy Bugle Boy." The girls are dressed in military outfits just like the Andrews Sister. The audience is full of patriotic civilians and military personnel from all branches of the service. Everyone is on their feet and many are dancing in whatever small place they can find. They didn't know it yet, but these were the people that decades later would be referred to as "The Greatest Generation." The girls' performance was scheduled to be one and a half hours, but after many encores they finished with a crowd favorite: "Rum and Coca Cola."

After the curtains are drawn, the girls come to get the men for a private party backstage. Schank says he loved the last song and asked if the gals have any rum and cola. Linda says, "Well, boys, it just so happens that we do."

The girls excuse themselves for a moment and retreat to their dressing room to get out of their military costumes. Barbara says, "Ok, here's the rule: one drink and we offer to hang out with them on the boardwalk." They return to the men with the drinks and say, "Let's enjoy these boys; it's all we have." Rose asks them if they had been in the service. Spider says, "Well, I flunked every medical for reasons I would rather not discuss and as for Schank here, well, need I say more?"

"Out of of the three of us I am the only one they wanted but I could not convince them I could shoot," Matt says, "Well, I guess I needed you there that day at the recruiter, Rose; we know now that together we shoot pretty well." They all break down with laughter.

They finish their cocktails and head for the Ferris wheel. After a short wait in line they all board the ride. The other girls are doing this to support Rose, who is clearly smitten with Matt. The ride handler says they are lucky, as this is the last ride of the evening since the lights-out order is soon. Along all the coasts of the United States there is a constant threat from German and Japanese submarines, so they enjoy this moment before the boardwalk is closed for the evening. Matt tells Rose of the time when he was a young boy and heard the Hindenburg fly overhead; they talk about what the world has become since, and Matt says that if he could find a way to take out Hitler he would do it himself.

"Well, Matt, I feel you are going to do great things with your life. I hope we can see each other again," and as she says the word "see" she gasps and says, "I am so sorry, Matt."

"Honey, we do see each other, and I am very sure we see things that others may not."

Tears of joy well up in her eyes and the couple embraces as they wait their turn to disembark from the ride. They are the last to get off and Spider is pacing and worried that they will get home far later than what they had promised. Barbara says, "Hey, I have an idea—let's get rid of these shoes and take a walk on the beach." The other ladies continue to be good sports as they are happy for Rose and Matt and thus are willing to endure more time with the other two men. The curfew is ignored and the group sneaks down a stairway behind the pier with the rides above and leaves their shoes hidden under some beach chairs that are stacked up against the seawall. Rose takes Matt's hand and the others follow down to the waterline.

The others lag behind to allow Matt and Rose to have some space.

"Matt, tell me about your family."

"Well, my parents came over on the boat from Palermo and were processed at Ellis Island. They came here with the equivalent of one hundred American dollars and a lot of hope. I have three wonderful sisters and a great brother. We all look out for each other and are very close. My mother and father are very strict and proud of their American citizenship. I work with my dad, who sells insurance. He works a lot with the Italian community and helped build a new church. He loves his work because it helps families protect what they have and gives him the opportunity to support our parish and our family with the money he makes.

"I go on sales calls with him and I have learned a lot about what it means to help others while earning a good living in a respected profession. Mother presses our suits every day so we look our best. Each day is started with a prayer, because, as dad says: 'You must have a servant's hearts to be a good salesman. You cannot sell anything unless you believe in your product and that it will be good for others. The best salesmen are good listeners, not those that just peddle something.' I often think that being blind has helped me to become a good salesman as I tune in to every word and motion that people make."

"Motion, Matt?"

"Yes, for example: I can tell many things about a person by the way they shake my hand. The manner in which a person shakes your hand can often tell if someone is sincere, confident, or even worried. The hands say so many things, from being smooth from lack of manual labor, weak from age, or weak from lack of conviction. When you took my hand I felt a warm heart, and the way you covered it with your other hand said you are a protector, like a blanket of sorts. What do you do, Rose?"

"Well, I am a protector, I guess; I am a nurse's aid. I work at the Naval hospital at the base not far from here. In fact my mother is a physical therapist and because of her I always wanted to do something in the health field. I grew up hearing stories of how she had helped people who often had lost hope in recovery. She never gives up on a patient and I have learned so much from her. Funny thing, though, Matt, is that I also like business and have thought about being an accountant. I know that is quite a change from what I am doing now, but I had an uncle who was in the profession. He was such a great man and let me earn some money by working in his office. I have a knack for working with numbers."

Matt and Rose bond quickly as they have realized that they have been raised to serve others and thus have immense respect for each other. It is getting very late and the others catch up to them on the beach. Spider insists that they must leave soon.

"Sure, for once it is you that wants to be the good mama's boy," Matt said.

Barbara and Linda chime in and say it is best that they return to the base soon; their shift begins in less than an hour. Matt and Rose embrace and exchange contact information. They walk back up to the boardwalk, where a police office sees them and says, "You folks had better get a move on. You should know better, the curfew started a while ago."

The six of them head for the parking lot and the boys offer the girls a ride; they refuse, as they are catching a bus soon with others from the base. Rose takes Matt by the hand and leads him away from the group.

"Matt, you be sure to get in touch with me soon, ok?"

"Yes, I will; you can count on it."

They embrace and she kisses him on the cheek. His smile has never been brighter and he is frozen in place as she walks away. Spider says, "Hey, snap out of it, lover boy." Matt doesn't say a word and the three head to the car. Matt gets in the passenger seat and Schank lays down in the backseat. Spider starts up the car and the three drive by the bus stop where the girls are boarding their bus. Spider says, "Hey, Matt, give a wave; they are to your right." Matt hears Rose call his name and the two exchange a smile and a wave. Matt is speechless and still not saying anything.

"What's wrong? Cat got your tongue?" Spider asks.

"Spider, just drive, ok?"

As the car heads down a dark road somewhere in farm country in Colts Neck, New Jersey, the car begins to make some grinding noises and begins to lose power.

"Damn it!" shouts Spider. "We are in the middle of nowhere and now we are breaking down."

"No worries, Spider, I am sure we can figure this out."

The men all get out and Matt leans against the car as the other two argue over various things that could be wrong with the car. They soon resolve themselves to the fact that they are stuck.

"Hey, do we have any hooch left in the trunk? We may as well finish it up; we are going to be in a world of trouble anyway."

The men proceed to get very drunk and decide to push the car.

"Hey, Matt, this is your chance to drive!"

"Hey, I can do this!"

Matt jumps into the driver seat. A few hundred yards later a local police officer pulls up alongside the car.

"You boys look like you could use some help."

"Yeah, more than you know; our driver is blind," Spider laughs.

The cop looks at Matt and he takes his hands off the wheel.

"Yep, that's true; those idiots back there thought this was my chance to drive."

"Well, now I have seen everything," the officer says. He sees all the empty beer bottles in the car and says they are lucky he is in a good mood; he tells them he will have a tow truck come out and bring the car to a local garage.

"In the meantime, you three clowns get in the back. I will bring you to the station where you can call someone for a ride."

Matt is awakened from his dream when they hit some large waves. Matt buzzes Damiano on the intercom and asks him to come below to the master suite.

Once he is in the room with the door closed Matt begins talking.

"Damiano, I am not sure how I am going to pursue this thing with Dawn, but I have some ideas. In the meantime, I want the locks changed on all the properties as well as any alarm codes to all of the residences except the condo in Jersey City, New Jersey. Please call that bitch and tell her that she can have access to that home only and that she may want to pull out her copy of the prenuptial agreement that she signed. I also want her tailed everywhere she goes, and remind security to review all videotapes from all the properties. As for work, call my assistant Cindy tomorrow and tell her to call me on the satellite phone as I will be joining my daughter in Italy along with the boys. Tell Cindy that she is not to tell Dawn any of my whereabouts. Damiano, I don't know what's worse: the booze or the waves. I have to focus and keep an edge on my situation. Damiano, go topside and help the guys I need to lay down for a while."

He braces himself against the bulkhead of the stateroom and makes his way to the to bathroom door. While walking he loses his balance in the rolling seas and hits his head on a shelf; he is knocked unconscious for several minutes. He regains consciousness and feels the warm rush of blood coming down his face; he calls out for Damiano. The boys and Damiano rush to the stateroom, where by this time, Matt has put on his game face and laughingly says he had a bit of a wrestling match with scotch and ocean swells.

"Head wounds tend to bleed a lot, so the mess look such worse than the actual cut," Damiano says. "Matt, don't worry; this is a very small cut. It's ok, just a small bandage will take care of this. My friend, I have known you a long time, and in the past forty-eight hours you have had more physical trauma than you have had in the last twenty years."

"Yeah, my friend, I need to get my wits about me and slow up on the booze; I have things that need my immediate attention and all this drinking is not helping. By the way, have you been able to reach the airport? Is our crew and plane going to be ready for our departure later today? I want to be with my kids, far away from here, as soon as possible."

Damiano says they have been busy with the chores topside, like storing fishing gear and washing down the decks.

"But I assure you, my friend, all will be taken care of promptly."

Damiano finishes patching Matt up, yet again, and turns to him as he leaves the stateroom.

"Hey, your fall could have been worse; you could have lost an eye."

Matt laughs and says, "Only you could get away with that one, Damiano. Hey, do me a favor: reassure the boys. I know they are worried, to say the least. They trust you and it would be good to hear words from you that make them feel at ease. I am sure everything is going to be ok. I have been thinking of a plan for dealing with Dawn and her boyfriend and will share it with you after I have spoken to the lawyers and the security team. When we get home today I want all the photos of Dawn removed, and please get rid of that mess of a painting hanging in the front hall. Have extra security detail secured at the house in New York City and here. As for Shamus, it's probably a good idea to have a tail put on him as well. I know my security folks are good, but order them to use new faces that Shamus and Dawn have never seen, and come to think of it, until we know who her lover is, it may be best to have you work closely my attorney Jeff and help him coordinate all of this."

"Yes, Matt, that's a great idea; I will take care of that."

"Damiano, I am going to stay here for a few minutes. I want to gather my thoughts before we get back."

"Sure, Matt, relax. We will help Billy get things battened down on the boat. The men all gather on the fishing deck and keep their voices very low. Damiano assures the boys that their father is very much in control of the situation and is looking forward to the trip to Italy.

"I know your father had mentioned not to speak of this to your sister, but it is imperative that there be no discussion of this anywhere," Damiano says.

"It's important that some others know, don't you think, Damiano?" Luke asks.

"No! Luke, under no circumstance is this to be mentioned again until your father gives further direction on this. Am I clear?"

The boys are stunned by Damiano's demeanor and glance at each other and concur in unison, "Yes, yes, for sure, Damiano."

"Ok, then please help Billy with this boat. I have to make a lot of calls to make and must speak to people at the airport to make sure the jet and crew are ready this afternoon."

Damiano opens the door to the helm and asks Billy how much longer until it will be until they are back at the dock. Billy says about a little bit over an hour.

"Ok, boys, we have a couple of hours to get this baby shipshape."

"Wait a second, we forgot one of our best traditions," Luke says. "Damiano, do we still have that revolver tucked down in the engine room?"

"Ah, yes, we do. Let me go get it."

On the way Damiano stops outside of the door to the master salon, where Matt is resting.

"Hey, my friend, the boys are asking for the revolver, as you all forgot the tradition."

"Sure, go get it, but humor them and make sure blanks are it in. You never know where those bullets will land. I will head up to the stern and meet you there."

Hey, boys, Damiano is getting the gun; that was a great idea. I can't believe we forgot. Are we lucky enough to have a day sky moon up there someplace?"

"Yes, dad, high in the sky to the East," Luke says.

"Ok, you know the deal: youngest to oldest."

Damiano approaches the fishing deck with gun in hand and Billy notices it. The two men exchange a nod and smile and Billy slows the yacht to a slow idle to keep the stern in position with the moon. Mark takes aim and shoots the moon, as does Luke.

"Hey, Damiano, lets go out of order this time and you go before me," Matt says.

"Ok, Matt, and he takes a shot and hands the gun over to Matt. Matt hesitates.

"You know, sometimes it is ok to break from tradition, so that is why I honored Damiano by letting him go before me. Damiano is family and I want you all to know that he is going to be family forever. So, boys, Damiano: this shot is for you, my friend . . . Hey, Damiano, you mind heading to get us all a drink? I want to make a toast before I take this shot."

Damiano is hesitant but complies. The boys offer to help and go with him. Matt goes to the edge of stern and empties the blanks into the water where no one will see this take place. He reaches into his pocket and replaces the blanks with real bullets that he had gotten from the nightstand before heading back up. The men return and Damiano hands Matt his drink. Matt places the gun in his right hand and the glass in his left. He says he will make his toast as he takes aim.

"You boys stay there by the chair; I want you to have a little space. This is going to be my best shot ever. Luke, go get that flag Dawn bought for the boat. You know the one that says "Dawn's Early Light"?Man, I am glad I did not name this boat that. Ok now hold it up outstretched in front of the barrel of this gun and tell me when my aim is right. Boys, this drink is to honor my long time friend, Damiano, who with this shot I declare a family member for life."

"Ok, dad, let it rip."

Matt pulls the trigger and a hole is blasted right in the middle of the flag. Damiano cannot believe his eyes.

"Men, you had blanks; I had a real bullets. Now, this was a test. If a blind man can pull one over on you it is time to raise your vigilance." The men are all stunned and speechless as Matt directs Billy to take the yacht back in and further lectures everyone to not forget the lessons of the day.

The sun is starting to get low in the western sky and the water has begun to settle down as the yacht gets closer to the coast. Matt asks Luke for the flag and he then tosses it overboard where is it consumed by the wake of the engines and disappears beneath the surface. The boys approach Matt and the three embrace as Damiano and Billy look on.

"This is a new chapter," Matt says, "A new beginning for all of us. Lets not be sad but rather recognize our blessings and get our butts on that plane as fast as we can."

As the yacht pulls into the marina they pass friends sitting on the deck of a rather large Hatteras yacht. They acknowledge the group with waves and shout that it's great to see them. Matt says, in his usual cheerful reply, "It's great to 'see' you, too," and under his breath he adds, "You arrogant lazy bums." The men all have a good laugh at that remark as Billy backs the yacht into its slip. Matt whispers to Damiano to make sure the gun is secured properly and that the live rounds of ammo are locked in a drawer in the master suite before they go. The men disembark, get into the Suburban, and drive off to the house to pack for their flight to Italy.

As the men approach the house and open the gate with the remote control the sound of the bell inside the house alerts the dogs to their arrival. The remote control in the Suburban sends a special signal, which offers a different ring tone so those in the home can distinguish the family vehicles from others. The dogs know this bell and begin to bark and get excited about their arrival and Damiano is asked to pull directly into the garage so that they may easily load the vehicle once packed. By now the dogs are all gathered at the door as they know Matt will be giving them a treat as he always does when he comes home.

"Hey, Damiano, I am going to hang out with the dogs for a bit. Can you gather my summer travel clothes? You know what I like. But whatever you do, please don't pack that hideous lime-green jacket that Dawn thought was so nice. In fact, please pack it up with the preppy stuff she has gotten me. I am fairly certain that you know the stuff I don't like, and you can take it to the goodwill when we are gone. You know how much I like my things from Guido, the tailor down in Palm Beach. In fact, that reminds me: we have friends staying in the Palm place for two more weeks. Please let security know to properly shut it down the day they leave. Dawn loves that place and I have a feeling she will try to head down there and I don't want her anywhere near it."

Matt grabs a handful of treats from a jar in the kitchen and heads out to the back porch with the dogs. They have a playful time and he tells them they will be missed and that Damiano will take good care of them so they, in return, should watch out for him.

"Damiano is getting older and we all need to be mindful of that," Matt says to the dogs.

The boys come out to join Matt and play with the dogs for a few minutes and they tell Matt that all is ready at the airport.

Chapter 6: DEPARTURE

Blake, their head pilot, has advised that he would like to leave within two hours to beat some weather that is moving in from the south. So with that they have a few more fun moments with the dogs and head back into the house, where Damiano is ready with their things. He asks the boys if they don't mind heading up to get their suitcases on their own as he is a bit out of breath from the long day and packing Matt up. Once the boys are out of listening range Matt asks Damiano if he would like to have someone stay with him while they are gone.

"Damiano, I think you could use some company; I know you will be worried. Feel free to have anyone you like and trust come stay here with you. I will be in touch frequently. Let our folks know of any communication you get from that witch."

The plane is kept at MacArthur Airport in Islip,

New York, which has the facility and runway to handle their converted Boeing 737. It has all the luxury of a living room, a couple of sleeping cabins, and an office and conference cabin. The men all pile into the Suburban and head down the Long Island Expressway. Matt is always excited to be anywhere near airplanes. When he was a young boy and still had his sight he would dream of flying one day. When the Hindenburg came over his head in Jersey City and he could only hear the sound of it passing it reminded him of how much he loved the thought of flight. He was never scared of it, even though things back then were not nearly as safe as today.

The men are all very excited about the prospects of the trip.

"I am sure we are going to put on some serious weight when we are there," Luke says. "Sis has been saying they never stop feeding her."

"I don't remember the people over there very much," says Mark. "Do you, Luke?"

"We were so young the last time we were there."

"Yep, Luke, you were nine, and Mark, you were six," Matt says. "Your mother had her hands full with you kids and all the relatives wanted was to teach her how to cook real Italian food."

The car phone rings and it is Blake, asking how far they are, as things were ready to go at the airport.

"All good, flyboy, we are pulling off the highway now," Damiano says, "See you in the hanger in five minutes."

As the Suburban approaches the hanger, Damiano presses a specially coded remote control for the large barn-like doors, which slide open, and they pull up next to the plane. Blake and his other two crew members are waiting for the men and Blake states that all the pre-flight checks have been completed and the plane is fueled and ready to go.

"We have a tight window here, guys. I will help you load up with Michelle, our flight attendant, for this hop while my first officer completes all the cross checks."

All are aboard in very short order. Michelle asks them if there is anything they would like as they are settled in. Matt says, "Yes, please, get us all some ice water; its been a long day so I am calling this one. We need a break from drinking for a bit."

They settle down into huge, comfortable, reclining lounge seats. The engines are fired up and Blake announces that all should be belted for taxi and that they have been cleared as number one for immediate departure. The taxi is very brief and the huge plane is up in the air in moments. Matt announces their flying time to Palermo will be eleven and a half hours and that it is expected to be a very smooth ride with some potential for weather once nearing their arrival.

The cabin of this plane is something extraordinary. Every inch of the plane's interior living space was customized to the specifications of the family's interior designer. The flow was carefully designed to have an area where people could relax and speak while others could break off for a meeting or rest in the plush sleeping cabins. As the plane ascends to cruising altitude Matt recalls that there is a family photo album in the master cabin.

"Michelle, once Blake gives the ok, please go back there and look in the cabinets above the desk in my office. There should be a photo album in there that the boys may not have seen for a while and I think there may be photos of some of the relatives we may be seeing on this trip."

"Yes, Mr. Messina, most certainly, and we are well stocked with some of your favorite foods so please let me know whenever you gentlemen would like to eat."

"Thanks, Michelle, will do."

"Ya know, dad, its weird that we haven't been back there for such a long time," Luke says.

"Yes, that's for sure; when you three were growing up we juggled an awful lot and lost the connection with family over there. Your sports and our businesses owned our lives. For a while there it seemed like your mother's full-time job was to organize your sports travel and practice schedules," Matt says. "I am so proud when I can tell people of your accomplishments. To have three kids that all progressed to be college athletes and scholars is confirmation that your mother and I raised you correctly. In fact, I am more proud of that than any business achievements you mother and I made along the way. Follow your dreams, boys, and everything falls into place over time."

Blake announces they are at cruise altitude and are free to relax as they wish. Michelle retrieves the photo album and presents it to Matt.

"Mr. Messina, here you are, sir, and if there is anything you gentlemen want please hit the pager. I will give you some privacy and head up to check on things up front."

She closes the living area door, passes to the flight deck, and gives a knock on the door. The first officer, Conner, opens it.

"Hey, good looking, what's up back there?"

"Keep the cracks to yourself, Conner. Blake, if there

is anything I can get you or Mr. Flirt here, just let me know. I am giving the Messinas some alone time back there. They are looking at a family album and I am sure they could use some space."

Mark takes the album in hand and turns to page one, where there is a picture of their family. The five of them are seated under a beautiful tree on a hillside in Palermo on a bright, sunny day. Mark is transfixed and unable to move as Luke leans in to look at the photo.

"Dad, the first picture is all of us. We look so happy in it."

"Yes, it was a wonderful time. We had spent that day eating an amazing meal that our relatives had made. We all participated in making the food, from baking the bread to crafting the tomato gravy. Your mother listened very carefully to my aunts as they dispensed the secret ingredients and loving process required to get the desired results."

They all smiled and Mark says, "Mom really got it; her gravy was like no other we have had in restaurants."

There is a long pause and Mark's voice is choked up as he tries to speak more about his mother. Matt tells him that it is all right and that he needs to let his emotions show because he never did when she became sick.

"To this day you have never shed a tear," Matt says.

"Dad, I had to be strong for you; we all had to be."

"Son, I love you for saying that, but never confuse the emotion of crying with weakness. People say things like, 'We need closure.' Let me tell you, boys: there is nothing more silly than the notion of closure. Life is a journey through time. We don't shut the door or close anything. Everything that happens in life is an experience. We learn and build from each person and experience in our lives. Mark, you have been holding things in since your mother died and that can eat at your soul. We love you and you need to let it out. At times you have even become angry at things that you should not be mad at. Holding things in can eat at one's very soul. This trip is about reconnecting for all of us. Gents, let us use this time for all of us to get closer and recalibrate. For me, I have a lot to be mad about, and that can be a perilous thing if left unchecked. Whenever I get mad at someone or something I take a step back and think about the big picture from which I develop a plan that is not reactionary but strategic and tactical. I am not very proud of some of the things I did at the house today, like ruining a picture of Dawn, but . . . it did make me feel good for a brief moment. I want to spend time with you guys and the family to begin the process of thinking about what I am going to do as the head of the family to keep us all together and at the same time move Dawn out of our lives in a way that is the most . . . well . . . let's say efficient."

The boys look at each other and Luke says, "Dad, we understand and will be there for you no matter what. Just tell us what we can do to help."

"I love you, boys, and lets take the time in Palermo to really feel the old country and our people," Matt replies. "Their hearts are bigger than life itself and they can teach us the real meaning of life, family, and what is needed for long-term happiness."

They continue to look through the album and describe to Matt what they are seeing. As they do, Matt slowly drifts into a sleep and dreams of Rose, their courtship, marriage, and a fateful honeymoon trip to Florida.

It is June, 1955. Matt and Rose decide to head down to the beach in New Jersey for the day. At this point in time, Matt has started a small insurance agency in New Brunswick on the first floor of their family home and Rose is working on getting her bachelor's degree and CPA from Rutgers University. Matt's first office is located in a the front of the first floor of the family home and is very spartan in its appointments. Matt is seated at a desk in the office and his sister, Stella, is just in front of him, positioned as the receptionist. Rose walks into the office and is greeted by Stella, who gives her a warm hug. Matt is on the phone helping a client whose home had just burned down. His words are very calming and assuring. The two women look at him and smile at each other. Matt tells the client that he will be over shortly to assess the damage and help them get on their feet. As Matt hangs up the phone Rose walks toward him. He stands and the two embrace and kiss.

"Well, honey, it looks like we have to make a stop on the way to the shore. One of my clients, the Armana family, just had a fire and I should be there to help them."

"Sure, Matt, let's do that."

"Stella, can you please write down the directions for Rose while I get my briefcase together? I already have my beach things in the bag behind my desk."

Matt packs some braille paper into his briefcase, along with his braille board and a metal device that looks like a slide rule of sorts that helps him guide the braille punch over the paper. The two head out to Rose's 1952 Chevy Bel Air. It is a beautiful car, light blue with white interior. Once in the car, Rose says, "I am pretty sure I know where this is. Over the bridge on route one into Milltown, correct, Matt?"

"Yes, it is the first left then right once over the bridge."

As the car gets close Matt is the first to pick up on the scent of smoke and then Rose. As they pull down the street, fire trucks are everywhere and neighbors are out on their lawns watching the drama play out. Rose rolls down the window to ask a police officer where the Armana family is, as she is here with the insurance man. The policeman looks in the car and recognizes Matt.

"Hey, Matt, how have you been? It's Lou, from church."

"Lou, my friend, I am great, but it I am worried about the Armanas. Where can I find them? I will take you and this pretty lady to see them. They are at a neighbor's house across the street."

"Lou, I want you to meet Rose; she is my girlfriend."

"Well, well, well, aren't you the lucky one, Matt."

The three head over to the house where the Armanas are. They knock on the door and Lou explains this is the insurance man, Matt, and his friend. They are here to help the Armanas. They are led into the kitchen where the family is being comforted by several neighbors.

"Mr. Messina, it is so good to see you," says Mr. Armana. "We are all safe but I am afraid the house may be a total loss."

"Not to worry, you have excellent coverage," Matt says.

"Matt, great seeing you; I have to help outside," Lou says.

Matt and Mr. Armana sit down at the table. Matt takes his braille set out and sets up the paper and braille guide to take notes. The others look on in amazement as he begins to ask a few questions on how they think the fire started.

"Well, I am sure it was the toaster. It started to spark the and next thing we knew the curtains were on fire. We focused on getting the kids out and when we went back the entire kitchen was on fire and was spreading fast."

Matt taps all of this into his braille paper and then pauses to tell them not to worry; all of this is covered by their insurance and will be paid quickly.

"I will get the police report from Lou in a few days; please call my office tomorrow. In the meantime, here is $500 to get your family into a hotel and buy some clothes."

"Matt, we can't take this from you."

"I will get it back from the insurance company as an advance against the claim. I have the authority to grant up to $500 for emergency claims like this."

The Armanas are overwhelmed and grateful. Rose is watching the others in the room carefully and realizes that what has just taken place was not only a great thing for the family but an amazing promotion for Matt and his agency. The Armanas and the couple embrace and are led back to the door. Rose leads Matt to the car and as they drive away the Armanas and other neighbors gather at the windows of the home to see them off.

The two drive off and make their way around the emergency vehicles and crowd.

"Honey, you are an amazing man, and I love what you do for a living."

"Yes, claims like this are a chance for us to serve and really be reminded of why we are in this business, besides making money. I love to help people in their time of need. "Yes, and I think something else happened back there."

"Yes, I know what you are going to say. I expect that many of those neighbors will be calling me for insurance."

"Yes, that is it exactly. You are in a great business. Helping people is such a noble cause, and to think you can make nice money while doing it."

"Thanks for coming with me, Rose; I am happy you could really see what I do. People think all we do is sell things and forget about the responsibility we have to others. My father instilled solid values in me and I hope that I can be half the man he is."

"Well, Matt, I am sure you will, and your father must be incredibly proud of you."

Rose comments on what a nice day it is, and that the drive should be good as the traffic seems light. The drive through the farm country of Colts Neck and surrounding area is beautiful. Rolling hills and large ranches with infinitely long white fences line the road. The drive goes by quickly and they arrive in just under one hour. As the two pull into the beach town where they met, Rose comments on how hard it is to believe that they have been dating so long.

"Yes, Rose, I am so glad that you have stuck it out with me. I have always felt it was important for me to get things on track for the business while you also secure your degree. I love this place, the way it sounds and smells, and most of all, that we met here."

They park the car and bring their bags to the public baths where they can change into their beach clothing then head down to the water. They lay a blanket down and place their things on it and go for a walk on the beach, hand in hand.

"Rose, I can hear that we are getting close to the amusement pier where we had that great first night."

"Yes, honey, it is close."

Matt stops and turns to Rose; he takes both hands and kisses her. He then drops to one knee and reaches into his pocket and pulls out a small box and hands it to her.

"Rose, will you marry me?"

She jumps up and down several times.

"Yes, yes, of course I will! I love you, Matt."

The two embrace and spin around on the sand and some passing by realize what has just taken place and applaud.

Matt's dream now takes him to the church where he and his brother are preparing in a small room behind the altar for the wedding ceremony. Isidore, Matt's younger brother, is a very handsome young man. He is slightly taller than Matt and built a bit more like a football player than Matt's baseball-player-like physique. He also goes by the name Izzy. The two are quite dapper. Isidore asks Matt if he is ready.

"Yes, of course, brother! I am the luckiest man in the world. Izzy, how do I look, am I all set? This is the first time I have had one of these monkey suits on.

"Yes, brother, is it very sharp and you look great and don't let uncle Nick hear you call that a monkey suit; he spent weeks tailoring these for us and the others in the wedding party."

The two walk toward the window, where their home can been seen just a few houses down from the church.

"I am so proud of you, brother," Isidore says. "I am happy for you and our family."

Matt places his hand on the window as if to "look" out and says, "I am grateful to have you has my brother, and our sisters are the best anyone could have."

"Yes, Matt, they are, but poor Rose will always be under the microscope as none of us want to let you go." The two exchange a few laughs. There is a knock at the door. It is an altar boy no older than twelve years asking them to join him and the Monsignor in his chambers. The three walk down a small hall resplendent with paintings of the apostles and Jesus Christ that lead to a huge wooden double door. The altar boy knocks and asks for permission to enter. The Monsignor opens the door and seats the two men at the desk; he then sits down behind it. The altar boy bows and kisses his ring and is excused from the meeting. The Monsignor asks Matt if he is prepared to receive the Holy Sacrament of Marriage and Matt says, "Yes, Monsignor I am in every way." The Monsignor is a very stern man, yet pauses, and with some emotion says, "My son, I baptized you in this very church as a tiny infant, then confirmed you as a young man, and I am very pleased that you are prepared for this Sacrament. I pray that I am here on earth long enough to baptize the children that god blesses you with."

The Monsignor stands and walks behind the two men, placing a hand on Matt's left shoulder as well as Isidore's left. He bows his head and prays for the health and happiness of the couple. The altar boy is called back in to lead the men to the altar. Music starts and the bridal procession begins. There is a pause in the music and the familiar music announcing that the bride is about to walk up the aisle is commenced. Matt and Isidore are like soldiers as they stand at attention in this most holy of moments. Isidore walks Matt to the center of the altar, where Rose is given to Matt by her father. The entire congregation, family, and friends helped prepare the church. The flowers are amazing and the choir was handpicked and supported by local professional singers as well. There is not an empty seat in all the pews and those that could not get a seat are waiting outside to get a glimpse of the popular couple and throw the traditional rice. The wedding goes flawlessly and ends with a kiss and they walk out of the church to the roar of applause. They exchange pleasantries with many of the people and then get into the limousine to take them to the Lakeside Manor, where the reception is to be held.

It is a beautiful day in the early fall. The leaves on the trees are just beginning to turn beautiful shades of red and yellow. There is a band playing period music. Matt asks Izzy to tell the bandleader to play "Boogie Woogie Bugle Boy" to get things into high gear and to commemorate the day he met Rose. As the music begins to play Rose teams up with her maids of honor. These are the same gals that were there and in the show the day she and Matt met. The three head up to the stage to sing together. No one is in their seats at this point and everyone begins to dance. The reception is all that a bride and groom could possibly hope for.

Soon, they are off to the airport to leave for their honeymoon and they arrive in Boca Raton, Florida, early the next morning. They love the beach and had been looking forward to this special time with great anticipation. They fell in love with the tiki bar, which was situated in the sand between the hotel and the ocean. They are at a beachfront tiki bar drinking from coconuts filled with Rum Runners dressed with cocktail umbrellas. A man wearing heavy gold chains appears from the beach and sits next to the two. He is a gregarious character wearing a T-shirt saying: "This is your lucky moment." He looks as though he has spent the greater part of his days in the sun: his skin was like tanned leather and what hair he had left on his balding head was sun and salt damaged. He leans over to ask the couple where they are from.

"Wow, before we tell you that, can you explain your shirt?" Rose says.

Matt asks what it says and then says, "Wow, I heard there were some interesting people down here: what is your name, sir, and why the shirt?"

"Well, 'sir'," he replies, "My name is Paul. I am visiting some investors in the area and I am from Key West."

Rose asks with a little laugh, "Investors? What is it that you do "Mr. Lucky?"

"Well . . . I am a treasure hunter."

" Ok, I think we are done here," Matt says. "Paul, would you mind leaving us alone; we are on our honeymoon."

"Sure, sorry that I seemed to upset you," Paul says, and walks away into the hotel. The bartender overhears this and walks over to the couple, who seem bothered by what just happened.

"Hey, sorry to interrupt you folks, but I just wanted you to know that Paul is a legend in South Florida and he and his team actually found some shipwrecks in the Florida Keys and that he was here for an investor meeting for a new venture."

"Well, well, well, that just beats all," Rose says. "Matt, I am intrigued by this."

"Rose, my love, I think we have had too many drinks to be levelheaded on this topic."

"Well, Matt, there was something about this man, and would you mind if we tracked him down to ask him more?"

"Don't worry, he holds his investor meetings here at the bar," the bartender says.

"Well, I rest my case; honey, lets go down to the water for a bit," Matt says.

The couple heads down to the ocean, holding hands; they walk into the water until it is chest high. They both enjoy the water immensely and stay in the water for nearly an hour in a loving embrace.

"The ocean will always be our special place, not only because we met there but because we are also spending our honeymoon here—"

"Rose, I know . . . you want to listen to Paul's pitch about treasure under the sea, don't you?"

"Matt, I have to admit I am very drawn to this weirdo's story and the apparent fact that he has some track record of finding things."

"All right, Rose, lets head back to the bar and have a few more drinks; hopefully we will get so bombed we will not hear a thing he says."

The two laugh and head back to the bar where more people were gathering. They grab the last two barstools left and quickly order up a couple more coconut-filled Rum Runners. No sooner have they taken their first sip than Paul reappears. He announces to all the bar guests that those interested in learning about his latest expedition should gather around by the picnic tables.

There are two tables situated next to the bar under a group of palm trees.

"Well, here we go, honey," says Rose, and takes Matt by the hand and they sit down with the others.

"Hey, well, it's the newly weds," Paul says. The rest of the people at the tables give them some applause and Paul offers a toast: "To your health and life, may there come many treasures."

Matt says, "Oh man, someone get me a double." Everyone laughs.

Paul opens the discussion and asks those who had already invested to raise their hands. About half of the dozen or so people eagerly raise their hands.

"Would those of you who have invested mind telling the others what you do for a living?"

The replies are impressive: civil engineer, historian, two divers, English professor, marine biologist, anthropologist, and an attorney.

"Folks, for the new people here I want you to understand that while many think I am off my rocker, those around that have invested already were handpicked and sought out by me. Each one of them conducted their own research on me and my theories about our new venture. Each member brings a skill set that this business venture needs and are passionate and committed to getting to our lucky day. I am still looking for folks that have money and faith in this team. We also need someone with an accounting background."

With that Rose smiles and she becomes giddy.

"Well, Paul, I am just finishing my degree and take the CPA exam in a month. I don't have experience, but this business can't be that complex."

The entire group laughs.

"Paul runs around doing deals on cocktail napkins," the lawyer says. "I am feeling like a babysitter, but frankly, I am all in on this. If you want to be his accountant, I am telling you now that he will be a nightmare for you as far as paperwork."

"Well, Paul, lets hear more; I am curious why you selected an English professor to be on the team."

"This will be an amazing story, and someone needs to archive things so that we can produce and book and hopefully a movie," Paul says.

"Oh my goodness, no wonder there is drinking at these meetings," Matt says.

Everyone laughs and the historian says, "That's not all; stick around and we will smoke some special cigarettes."

Matt stands up and says, "That's it, we are out," and asks Rose to leave with him.

"Oh, honey, please let's hear what's been happening here."

As the two talk more about leaving Paul opens a briefcase and says, "This is what we have found in the last two weeks." A few dozen gold coins and pieces of pottery are placed on the picnic table. There are gasps around the table. In a guarded whisper someone says, "We should be doing this inside in private room."

"No, I want to cause a commotion to get the word out to get more investors," Paul says. He explains that he thinks the ship, a galleon named the Alucia, had broken up in shallow water on coral causing the ship's hull to have tears, causing things to fall out over a large area.

"This is why we have been finding things over a three mile path so far. I want our historian to go to Portugal to research archives and manifests. We need to know more about the path of the Alucia. If we can find documentation, that will help us in our search.

"That may be a good idea, as the Portuguese were known for keeping great records," the historian says. He further suggests that he and his wife will make a vacation out of it and head there.

"Honey, these people are serious," Matt says, "We both have very little savings and my business is just starting to grow. This is not for us."

"Matt, please, let me ask my mother to lend me the money," Rose says. She asks how much is needed to buy in.

"We have created a core group with preferred stock," the lawyer says. "You can get the last of it for $25,000 and that completes our first round of funding. Each one of us has invested that amount. We are planning a second-round offering of a different class of shares for future investors, but we all have to agree that a start-up accountant is what we want."

Paul jumps in and says, "Let's give her chance. I have a feeling she will be great, not to mention we have been searching for a while and we have to get our ship in order."

"Matt, don't worry, I will ask my mother for it."

"Rose, that is entirely up to you; this is all too much for me and my risk tolerance but if this is something you feel good about I will support it."

Paul calls for a vote, "If the newlyweds can raise the funds for their share, do I see any objection to that or to Rose becoming our bean counter?"

They all weigh in on her lack of experience but love her attitude and it's settled: Rose is on the team pending receipt of their investment funds from her mother.

The group is now complete and the next topic of the meeting switches to organization of a budget for the dive team and the maintenance of the one, rickety old boat that they are using to explore an area just off Key West. They also plan an allowance for Dr. Lawrence, the historian's, trip to Portugal. Dr. Lawrence is a professor emeritus at Princeton University and has extensive writings in the area of archives and record keeping. He tells the group of a list of libraries and other locations where he will be going to research as much as possible on the history of the Alucia.

"I assume you have a name for this organization of investors?" Matt asks.

"Yes, indeed we do," the English professor replies, "We kicked around lots of ideas on that front and settled on Intrepid Expeditions."

The group's enthusiasm increases with the amount of alcohol consumed. Matt is clearly the most conservative in the group and goes along with it. His love for Rose is immense and he trusts her judgment.

"Well, honey, if these people actually find the

treasure this will set us up for life and you can have a built-in PhD thesis if you go on to graduate school."

The drinks continue to flow and the bartender sends over a round on the house. At this point a band shows up and the party kicks into high gear with steel drum calypso music that can be heard from all directions on the beach. Soon beachgoers are attracted to the scene and the place ramps up into full swing. As the sun begins to go down Rose remembers that they have dinner reservations planned at a waterfront place that the concierge had recommended. The couple finds the group in the crowd and says that they will be leaving for dinner; they ask when the next meeting is and exchange contact information.

"My friend, I may be blind, but I see more than you can imagine," Matt whispers in Paul's ear. "I really don't like any of this. Me and my family stick together and we will protect Rose. Don't mess with us."

"It's going to be fine, Matt. I know we seem like a bunch of drinking dreamers, and that is because we are; however, we do know what we are doing."

"You had better, Paul . . . you had better."

Rose asks Matt what he just said.

"Not to worry, Rose, I just told him I am cautious."

She gives Matt a long kiss and says, "I have a feeling this investment is going to be wonderful. I truly love you, Matt."

They head back to the hotel room and Matt relaxes on the balcony, which overlooks the party on the beach, as Rose takes a shower. Paul can see Matt from where he is standing. Matt senses where Paul is from the distant sounds of his voice. Rose shouts from the shower that she is done, and with this Matt stands up and points a finger at Paul. It is as if Paul saw a ghost as Matt turns around and walks through the curtains into the hotel room.

The newlyweds arrive at the restaurant, beautifully dressed in tropical garb. They are welcomed to area's best waterfront dining establishment. Many look on as the two are escorted to their table. The couple usually attract a little more attention, because most people are curious when they recognize Matt's blindness—not to mention that they are a beautiful young couple. They are seated at a table right on the water and order a bottle of champagne. Rose is quiet and stares at Matt.

"Why the silence?" he asks.

"I cannot believe how happy I am and how much I love you. You are so handsome, Matt; I just can't help but look at you."

"Me too, honey."

They both get a bit of a laugh at that and reach out to hold hands.

"What a day this has been my love," says Rose.

"Don't worry, honey, my mother will give us the money for the Alucia investment."

"I am not worried for us, but to risk your mother's hard-earned savings on something like this . . ."

"Well, mom is more adventurous than you think, love."

They have a great dinner and dance the evening away. The two wake up in their room and Matt turns to Rose.

"Hey, I had this crazy dream that we met some treasure hunters on the beach and your mother gave them money . . ."

They have a playful pillow fight and order room service. The morning is beautiful; as the sun rises they head out to the balcony in their robes and listen to the sound of the waves.

"You know, Matt, I could listen to that forever."

"Oh, yes, for sure, it is the pulse of the earth and I have my own theory on this. I believe, since the human body is largely water and most of the earth is covered with it, that is why we are naturally drawn to it. The waves and tides are much like the human ebb and flow of emotion."

Rose is taken back by the beauty of that statement and she stares at her love.

"That is a wonderful thought, Matt."

"So, Rose, now that we haven't had any drinks, how do you feel about this treasure stuff now?"

"I still want to speak to my mother, Matt."

"Well, Ok, I suggest you do that face-to-face. If we call her from down here I don't think it will be a good thing. She will be convinced we had been drinking and lost our sense of good judgment."

Chapter 7:

Adventure Begins

Matt's dream now takes him to the Newark Airport, where they have arrived back from their honeymoon. The couple is met at the gate by Rose's mom, Olga.

"I hope you two had a most wonderful time!"

"Mother it was beyond our expectations; it was utterly wonderful."

"You must be tired from a long travel day. I prepared extra food for dinner; would you like to stop on the way to your place?"

Matt and Olga have a great relationship. He is happy to accept the offer on the couple's behalf and is quick to tell her that while Florida has wonderful restaurants, they cannot replace the greatness of her home cooking and hospitality. The three head to the parking deck, where they load the car, and Olga says, "You two love birds jump in the back and relax; I will give you the limo treatment."

"Thanks, mom, that is very kind of you, but one of us will ride up front."

"Honey, I insist; it is my pleasure."

The couple slides into the backseat and hold hands. As Olga backs up the car she asks if they had any new favorite places down in Boca Raton and if they liked it good enough to return.

"Mom, we loved it and met some new friends."

Matt clears his throat and squeezes Rose's hand as to alert her not to discuss the Alucia folks as yet.

"Really, honey, new friends in such a short time?"

"Mom you know me and Matt; we tend to make friends everywhere we go."

"Yes, that is a nice trait the both of you have."

Soon they are on the New Jersey turnpike en route to Olga's home in the bucolic town of Westfield. The ride is short, and in less than a half hour they pull into Olga's driveway.

"Olga, this is so thoughtful of you to do this; we are wiped out," Matt says in a weary tone.

As they approach the porch Olga mentions that she has all the wedding gifts organized for them in the living room, most of which were checks and a dozen or so boxes. She says most of the boxes are from Rose's side of the family.

"That is not a surprise, Rose; it is a Sicilian tradition to give cash to support the newly married," Matt says.

Rose is excited to see what they received but says, "We can go through all this after dinner. Mom, is there something we can do to help you with dinner?" Olga doesn't reply. Rose heads into the kitchen to ask again while Matt waits behind on the couch.

Olga is bent over the sink, crying, "Your father would have loved this more than anything. I miss him terribly, Rose. It has been almost five years since he died and I simply cannot get over it. The wedding was so wonderful and to see the two of you so happy brings up memories of your father."

"Mom, I am sure he is with us more than you realize."

"Well, honey, you have always been a spiritual soul and I trust and pray you are right."

Rose shouts from the kitchen, "Matt, honey, can I bring you something to drink? We should have the dinner ready shortly."

"Yes, please, I would love a Coke if you have one."

Rose grabs a cold bottle, pops it open, and brings it to Matt.

"Honey, sorry, mom is emotional in there remembering dad so I will help her finish preparing dinner."

"That is fine, and please let's not bring up the Alucia folks now, especially with her state of mind; we really haven't given this enough thought."

"Certainly, love, I will be back in a few minutes."

Soon the scent of homemade pierogies fills the air in the house.

"Hey, you guys are really making me hungry in here; it smells great!"

"We are nearly done making our Polish version of ravioli."

The women come into the dining room with trays full of different types of the delicacy. Some are filled with cheese and others meat. There is even one type filled with sour cream and sauerkraut . They gather around the table, hold hands, and Matt gives a prayer of thanks. Rose dishes out a taste of each type to each of them.

"So, kids, tell me about this trip and your new friends."

"Well, mother, the beach and hotel were amazing. We spent much of it relaxing in the pool and ocean. There was this cute tiki bar on the beach that had steel drums and other types of music. We spent every evening there dancing—and of course drinking tropical drinks. We met some locals there and may visit them on our next trip."

"Oh that sounds great, Rose."

Nothing is mentioned of the Alucia or those associated with the expedition. Olga serves coffee and asks if they have room for her homemade blueberry pie. Matt is quick to say, "There is always room, but do you mind if we take a break and relax a bit?"

"Sure, why don't we go back into the living room and go through your gifts?"

"That is a great idea, Olga; Rose, lets count those checks first!" Matt laughs.

"Sure, honey."

There is a healthy pile of envelopes that Olga has placed in a basket and the couple counts $21,000. They are overwhelmed with the generosity of their friends and relatives. In the boxes was an array of nice silverware in the pattern they had selected at the bridal registry.

"Well, Olga, I think we have worked up a dessert

appetite and I am ready for your pie now. How about you, Rose?"

"Oh yes, I can't resist the scent of it any longer."

They enjoy the warm dessert and Rose helps her mom clean up.

"Mom, we can't thank you enough, but I thinks it's time we head home; we are really tired from the long day." They try to fit all the boxes in the car but cannot with all with the luggage they had brought so some are left behind and all the checks are placed into one of their bags. Olga drives the couple home and helps them bring in their luggage and gifts.

"Thanks so much, mom; Matt and I can't thank you enough."

"Yes, Olga, we love you and will be back to your home in a few days once we get settled—even better, let's have you over for dinner this weekend."

"That would be great, Matt, thanks."

"They wave as Olga pulls away; holding hands they walk into their home and kick off their shoes and relax on the couch in an embrace.

"Matt, I cannot believe the generosity of all our family and friends! In fact, I am almost embarrassed to have all this money."

"I understand, but accept it as a tradition and we will do the same for others, Rose."

There is a pause and Rose says, "Hey, wait a minute; I just thought of something. We don't have to ask my mother for the money any longer. We have enough savings to cover the difference on the $25,000 for the Alucia investment and will have plenty left over to have a nice cushion in the bank. We have our house now and all we need to do is put away enough money to cover emergencies. Do you feel your insurance business is going well enough to cover the expenses if I work on the Alucia project?"

"Rose, I do, but, wow, this is crazy."

"Matt, we are young and this is the perfect time to take a risk. We don't have any children yet and the house is paid for thanks to all your years of saving from the insurance business."

"Rose, I really think this is something we should not do, but I feel that your instincts are good—after all you married me."

The two laugh and have another playful exchange with the couch pillows.

"All right then, Matt, it is settled. I will call Paul and tell him we are all in on this."

"Rose, there is one thing. I don't want you spending time down there without me and I have to be in the office. How is this going to work?"

"Honey, accounting can be done from anywhere and we can go down every six to eight weeks for a long weekend to monitor what is going on in person."

"Rose, I would prefer to keep this to ourselves. My family would not support this and since my brother is now in my business as a partner I owe it to him to be around. I don't want to lose my focus on the business; after all, it will be the thing that supports us and other family members now and in the future. It is not fair to my brother for me to get involved in anything that will be a distraction."

"I understand, Matt; let's make this clear to the others that we can support the Alucia team but have to keep our lives in balance."

"Rose, you should be in the sales business, not accounting; you have a way with things."

"Matt, we are a pair in that regard. I am so excited; let's call Paul now!"

"Honey, I think I need a coffee, if you don't mind. Would you please get me one first, Rose?"

"Sure, I will make us both a fresh pot."

The two settle into the kitchen and enjoy the fresh coffee. Rose reaches for the phone and pauses to look at Matt. He senses her hesitation and faces her and says, "Honey, it's ok. I am good with this, love; go ahead and dial."

The sound of the dialer seems to be louder than usual and fills the small kitchen with a pulsing rattle as each number is turned. There is a distant hello on the other end.

"Paul, is that you? I can barely hear you; this is Rose, from New Jersey. You met me and my husband the other day."

"Oh yeah, of course, Rose, how you be up there?"

"Well, we are great and have made the decision to invest and support the team with the accounting services."

"Well, well, well, that is just great, Rose. When can you get down here?"

"Paul, I must be very direct with you. Matt has a business as well as a partner so he needs to be up here and I will remain here to support him and finish things at Rutgers next month. I have to focus on the CPA exam but feel that I can do the paperwork from here and visit you every six to eight weeks. Would that work for you and the others?"

"Yes, Rose, I am sure that will be fine. In fact, the

team spoke as a group after you left and we were all in favor of you and Matt's participation. I can assure you there is no hesitation."

"Great, Paul, please have the attorney send us the paperwork and I will prepare a list of things I will need from you for the tax and accounting work."

"Ok, Rose, send my best to Matt and speak to you soon. I know I speak for everyone that we are excited that you have made the decision to join us. They are a very passionate group and I will do everything possible to make this a huge win for all of us. Rose, I am sure you realize that I may not be the best businessman but I am sure we are on the right track and will find the Alucia. We need folks like you to babysit me and my antics. I have been known to do deals on cocktail napkins in bars but I assure you I can get organized with more experienced people around me."

"Paul, I have only known you for a week or so but you will not change. I can't be a babysitter but will act as your guardrail. Just be sure to keep everything you do in a file each week and send it to me, but no more deals without the approval of the group, ok?"

"Rose, you are a tough one and I need that. I will behave myself—after all, you are married to a Sicilian from New Jersey. I had better watch myself!"

Matt overhears this and smiles and gives a thumbs up.

"Paul, you are quite the character. I will speak to you soon; take good care."

As Matt drifts further into his dream he is thinking of an early morning on the docks at the end of Duval Street in Key West. Paul is sitting on a pier post smoking a cigarette and drinking a cup of coffee and the crew is slowly making their way to the boat ramp. The stench of the fishing boats in this part of the Marina is rancid and the line of kingfishers and pelicans looks as though the birds are in a military lineup, waiting to strike for any morsel. The crew is clearly hungover; each carries a duffle bag that is loaded with personal equipment.

"Well, men, this is your lucky day; we secured the last of the required A class shared investors. It is a husband and wife from New Jersey. The wife, Rose, is an accountant who will oversee our operation."

Lance, the eldest and most senior diver, says, "You mean over-sea! We have been getting by on adrenaline and rum and now we have to deal with a bean counter?"

"Well, Lance, that is true, but we have to become a disciplined organization if we are to win at this game. We have two well-capitalized competitors out there looking for the Alucia and they have far more cash and backing than us. Our boat is barely floating and we are rigging our own sand-sifting devices and underwater metal detectors. Soon we will have all the best equipment and be able to compete but I will tell you our real secret weapon is Dr. Lawrence. He is heading to Portugal to research the archives of the Alucia and its related manifests. He will be there for at least a year on a sabbatical from Princeton."

"Just great, an intellectual and a dream; lets get on with it today and head out to the reef off Dry Tortuga, boss," Lance says. "This has been our most productive location."

Paul agrees and the others board the rusty, old boat. The scene is a mess: a ragtag group of hungover drunks with three-day beards on a boat that is spewing smoke from a tired old diesel. The thirty-seven-foot old scow leaves the marina in a shroud of steam and smoke and heads out into the inlet for a day of diving.

Now, in his dream, Matt is working in his business office at his desk as Rose walks in. Matt's sister Stella is seated in the front as receptionist.

"Hey Stella, what is my honey up to?"

"Hi dear, he is on the phone with a client and I think he will be off shortly."

"Ok, I will head in there and wait for him."

The two exchange a warm hug and Rose walks down the short hall to Matt's desk, where he is just finishing his call. As he hangs up she walks over to give him a kiss; he knew it was her before she said a word from the scent of her perfume and the gait of her step.

"Matt, I have the papers from Paul and had a friend in the legal department at Rutgers review them. They seem to be in order; did you want to have our attorney take a look as well?"

"Sure, I think a second look would serve us well; our guy owes me a favor. I will call him now and offer to take him to lunch at Nardinis; he loves that place."

The couple head over and take their usual table and are soon joined by Jeff Heilmen, the family's trusted lawyer for years.

"Jeff, we have something for you to look over, but before you do we want you to know that we have not lost our minds."

"I don't think I like the sound of this, but sure."

They order a bottle of Chianti for the table as Jeff looks the document over. His eyes tell the story. Jeff drinks his wine like it's water.

"Please pour me another glass, because you surely must have had a few when you decided to do this."

"As a matter of fact we did, but we have thought about it for a while now and there are several reputable folks on the team doing this as well," Matt says.

"Do the two of you know any of them?"

"No, but the closest is Dr. Lawrence of Princeton University, and I had him checked out."

"I know of him as well. I have met him at several area fundraisers and social events. He has a very good reputation. What is his role with this?"

Rose explains to Jeff that all the A class investors have been selected because they have a certain skill set and Dr. Lawrence is researching the ships history.

"In fact, he is currently in Portugal, which is the point of origin and registry of the Alucia."

"Oh, I see; well, this is a fairly standard agreement. I am going to suggest that there be an accepted valuation method used to determine economic as well as intrinsic value if this stuff is actually found."

"Well, Jeff, that is an issue, because there are no accounting principles established for that."

"I see. In my professional judgment, I think you

are throwing money away. I can't add much to this document but I wish you nothing but the best."

The three toast to the success of the expedition.

"One last thing: do you know who represents the group from a legal perspective?"

"Yes, he is an investor as well and is handling all the legal work for the group. He is lives and practices in Florida."

"You guys mind if I call him and just get the lay of land a bit more before you send this back to him?"

"Jeff, can we call from your office? I don't want the others in my insurance office to know, and frankly don't want to give anyone the impression this will take me away from work. Frankly, Jeff, I don't want them to think that we have lost our marbles."

Back in his office, Jeff has a short conversation with the other attorney.

"It seems that everything is in order with the construct of the deal but you will have your hands full with this Paul character."

Matt agrees, as does Rose, but they sign the papers and head back to Matt's office. Rose pulls up out front and they kiss in the car.

"See you at five, ok?"

In his typical manner, Matt gets out of the car and walks up to the door without the aid of a cane. He knows how many steps to the door and his hand raises just as he approaches the door; he turns and waves to Rose.

As Matt drifts further into his dream, Dr. Lawrence is at a large wooden table in a dimly lit basement of a very old stone building in Portugal. There are very large books open and spread out over the surface of the table. He stands up and walks back and forth from one end of the table to the other, taking notes from each of the books. He seems puzzled by some of the common phrases in the archives that are not familiar to him. He takes copious notes and spends an inordinate amount of time on one book in particular. He is alone and begins to speak to himself. This becomes irritating to the curator, who is not far away, and asks him to please keep quiet as others are nearby trying to read and research.

"Yes, of course I'm sorry."

He picks up the large book to which he was transfixed and retreats to a more secluded area of the cold, old structure.

"I must call Paul as soon as possible," he says to himself.

As morning breaks he is quick to call Paul and tell him of some things in a book that have him confused.

"Paul, I have found some common things in several journals. First, I found certain words and phrases in the records of the expedition that tried to rescue the Alucia."

"Hold on, Lawrence, rescue?"

"Yes, another ship was sent to rescue or retrieve what was left of the Alucia and it too had been lost. In fact, they had located the mast of the Alucia, which was still sticking out of the surface of the water, when the recovery ship encountered trouble and that expedition was lost as well. A few survivors of the rescue took refuge in lower Palm Key."

"Hold on there, Lawrence; that tells me that the Alucia was likely in less than seventy-five feet of water, or perhaps even less, and may have been hung up on a reef off Palm Key. If that mast was still visible it was shallow. Please don't tell anyone of this. I am quietly moving our search to Palm Key today and will not tell the crew until we are out of the inlet."

"Ok, Paul, but before you hang up you may be interested to know a bit more about the manifest. I think that the Portuguese may have been smuggling emeralds out or Peru but cannot be certain. It seemed the

Portuguese may have been planning to finance something politically motivated with the emeralds but I cannot be sure. The translation is rough, but one thing is for sure—there were hundreds if not thousands of silver bars as well as countless gold silver coins on that ship. I am confident that there is more, given the fact that some rather wealthy families had commissioned the ship to do some trading."

"Lawrence, excuse the pun, but you are a gem. I am heading to the docks now."

At this time, back in New Jersey, Rose is opening a box of documents that Paul had mailed. She cannot believe her eyes. In the box are past due bills, letters from debt collectors, and cancelled insurance papers for nonpayment of the premium on the expedition boat.

"Oh my goodness, what have I gotten us into? Matt is going to be so upset."

She is in a near state of panic and calls Paul.

"What is this mess you sent me, Paul?"

"Relax, Rose, we have enough money to get us through another six months."

"Well, Paul, you clearly don't know the mess we are in. I have prepared statements for the investors. I would suggest we meet right away by conference call.

We have less than three months to go before we can't make payroll and catch up on all the bills, not to mention get the proper insurance for the boat and workman's compensation protection for the divers."

"Ok, Rose, calm down. I have some good news. Dr. Lawrence located some key information for us and I am moving the team to a different location. I am feeling very good about this new information. We have solid details on location from the ship's manifest."

"Wow, that is great news, but even so, I would suggest we raise some new capital, and to do so in three months will be nothing short of a miracle."

"That's ok, Rose, today will be our lucky day."

"Paul you are really something. Ok, take care, be well."

Chapter 8:
TREASURE

This day Paul takes the team in an entirely different direction once out of the inlet.

"Hey, boss, why are we heading northwest? What the heck are you doing?"

"Trust me, guys, I have some new information from Dr. Lawrence, but did not want to tell you drunks before we left the dock. We will be out for at least three days this time, so hunker down."

Where are we headed, boss?"

"Down to the Rojama reef off Palm Key."

"Wow, that is at least fifteen miles from where we located some of the chains. That is crazy."

"Gents, what I think may have happened is that the Alucia may have ripped her hull over where we were looking and drifted into Rojama in the storm where she got hung up."

Within an hour the ragtag bunch is anchored off the reef and in the water. Nothing is found the first two days. Paul is beginning to think that his theory is not correct and paces back and forth at the stern. He sees bubbles from a diver coming to the surface and then a hand breaks the surface with a large piece of pottery. Paul says, “Bring that to me!” It is clearly something that is old but he is not sure what it is. The others surface empty-handed. It is late and the sun is going down.

“Guys, I am very encouraged by this.”

They laugh at Paul and say it’s just an old piece of junk.

“You guys get some sleep; we are all going to dive tomorrow, and let’s make this our best effort as we have to head back soon.”

They are up at the first sign of the sun and set up a dive grid. They will cover a large area on this day. Paul is the first in the water and teams of two men take turns. It is around noon that day when Blake, Paul’s son, comes up with the largest gold chain they have ever seen. The team is elated and a few rush to get cameras to record the moment. It is six feet long and the links are huge. Paul explains that the links were used as a form of monetary exchange.

“Boys, we must be close. Look for anything that

looks weird and not like the typical coral formations you are used to seeing."

Pruit, one of the more seasoned of the divers, says he thinks he had seen something that seemed too symmetrical off the port bow about one hundred yards out. Paul orders him to take some photos of the anomaly.

"Blake and Pruit, you come with me. You others stay quiet on the radio; don't call on the radio or alert anyone about this chain."

The three head over to the area Pruit had described and in less than forty feet of water there it was: piles of silver bars and ballast stones from the ship's hulls all over the sea floor. The men are so excited they almost lose control of themselves. They open a dye locator to alert the team on the boat. They surface and are screaming to get some buoy locators over there. They had found the mother lode! The men cannot believe their eyes. In the shallows lay countless millions on the ocean floor. They bring what they can to the surface and realize that they should photograph the area before more is taken, as this will be used for archives to stake a claim as well as to detail the site from an archeological perspective.

The site is beyond description. Silver bars are in piles everywhere. These were the geometrical shapes that had been seen earlier in the day but not readily recognizable to the untrained eye as hundreds of years of coral and other growth had taken its toll on the artifacts. They begin to clear sand and before long a cannon is located. At the base is a clear marking that confirms this is the Alucia. The men surface and board the boat, where the rum is already flowing. Paul's wife was back in the office, as usual. It is in an old warehouse not far from the docks. The ship to shore radio is calling her in the next room. She acknowledges Paul and asks what's up today out there and she hears the phrase: "We will need more oxygen tanks by five." She knows that is the code for when they locate something very important.

"Paul, I repeat: do you need more oxygen tanks by five?"

"Yes, that is correct."

"Ok, I will meet you at the dock."

This code had been established so that others monitoring the frequencies would not be able to ascertain what was going on. In fact, they changed the code on a weekly basis so others outside the expedition would not catch on to any trends. Paul leaves with Pruit and a few others in the tender; the remaining team stays on the boat to guard their stake. As Paul and his men pull alongside the dock, his wife, Diane, assists the tie up and jumps on board and rushes to the helm and closes the door behind her. The two stare at each other for a moment, and with a huge shout of "We did it!" Paul hugs her, lifting her off her feet.

"We have her, we have her, and have confirmed it with a cannon marking!"

The two kiss more times than could be counted and tears of joy flow from their eyes. There is a state of shock and exuberance. Paul asks her if anyone knew back in the office or elsewhere.

"No, of course not, I did as we planned and have been very careful not to look excited on the way here from the office," she says. "Paul, do you have any samples with you now?"

Paul orders the others to get the boat organized.

"Don't do anything to call attention to us."

Diane and Paul go below deck. On the floor is a large blanket and Paul lifts it away to reveal a dozen silver bars and hundreds of gold and silver coins. She gasps at the sight.

"That is nothing; this is a very small sample of what we have down there," Paul says.

Diane drops to her knees and places her hands

over her mouth and begins to cry uncontrollably. Paul kneels behind her and places his arms around her.

"Yes, honey, this is the real thing. It's hard to believe the day is here; please take these coordinates and bring them back to the office and call the lawyer and Rose first. We will need to be very careful and diligent as to how we tag and process all this, not to mention keep the poachers out."

"Paul, I am more concerned about what that State of Florida will do with the claim, or perhaps even the United States or Portugal."

"Yes, Diane, I think it's all going to hit the fan and we need Rose and our attorney, Gordon, out in front of all this. I am heading back to get the boys; the sun will be going down soon."

Diane helps to untie the boat and heads to the office with a concealed grin as she passes all the familiar boats docked at the marina and the locals sitting on the stools of the outdoor bars. Many recognize her as she passes and she replies with her usual pretty smile and gentle wave. When she gets back to the office in the old warehouse, she closes the door and screams, "Yes! Yes! Yes!" Her private office is in a loft, which is accessed up a spiral staircase with metal grate steps. She can barely dial the phone as her hands are trembling from the excitement. Gordon is the first to be called.

"Hello, Gordon, this is Diane and this IS your lucky day."

"Huh, Diane, is this what I think it is?"

"Yes this IS!"

"Oh my god, I will be right there; please wait for me."

Rose is the next to be called and the lucky day phrase is repeated.

"Yes, Diane, it is," and she laughs a bit. "Don't tell me, hmmm . . . Paul actually decided to keep better track of paperwork?"

"No, honey, this really is your lucky day."

Rose holds the phone receiver away from her head as she recognizes the code phrase. She drops the phone and it smacks against the wall dangling from its chord. She slowly reaches for the phone and raises it carefully to her ear.

"Diane, are you telling me this is really our lucky day?"

"Yes, Rose, Paul and the team will need you here immediately."

Rose nervously gathers up her purse, car keys, and her briefcase and heads for Matt's office. Once there she finds it is very busy; there had been a storm the day before and everyone was on the phones helping clients with claims. Matt's waiting area was full of anxious people trying to get help with their claims process as well. Rose is recognized by a few of the clients so she stops to greet them and tell them not to worry, Matt's team will take care of everything. Stella is serving coffee and donuts to many of them. The Messinas and their staff treat clients like family.

Rose makes her way back to Matt's office, where he is on the phone, and she closes the door behind her. As usual, he knows it is her and gives a smile as she sits down and patiently waits for Matt to finish the call. Once off the call Matt stands to greet her.

"Matt, my dear, I see you are very busy here, but I have some news that could not wait."

Matt, thinking to himself for a moment, places his hand on her stomach and asks "Honey, are you?"

"No, Matt, but I am sure God will bless us that way. I came to tell you that today is our lucky day."

"Why, Rose, did we win something?"

"Matt, you know, 'lucky day'."

"Oh my, is this what I am thinking it is? Did the group actually—"

Yes, Matt! Paul called me a little while ago and I had to drive over to tell you in person."

"Rose, Rose . . . I don't know what to say . . . I am in a state of shock right now. Do we know any details about what has been found?"

"No, but Paul wants us there right away."

"Oh my, honey, I cannot leave; it will take us over a week to process all these claims. I am going out with my brother to visit many of the claim sites and I cannot leave him now. Rose, you understand, don't you honey?"

"Yes, love, I do."

"Rose, please book a flight once you get back home and don't tell anyone of this. Please call me back in a little while to let me know what the plan is."

The two hug and kiss. Rose leaves and closes the door behind her. Matt walks to the window at the back of his office and places both hands on the glass and raises he head up to the sky. To himself he says, "Dear Lord, please protect my beloved wife and keep us humble throughout whatever is to become of this."

Matt settles back down to his work and the pile of

files that had been placed on his desk. His claims processor, Marge, is paged to come in and read the notes to him. He then starts to make notes in braille from her readings.

"Sir, I want you to know the office staff has all spoken and we will do what it takes to help all these people; we already have over two hundred claim forms completed. If we have to stay overtime we are prepared to do it and we will do it without asking for more pay."

"Dear Lord, we are so blessed to have a team like this. Thank you, Marge, that is a wonderful offer, but Izzy and I have enough in reserve for emergencies like this."

"Ok, lets get to it then."

A short while later Rose calls Matt to tell him she can be on a flight the next day and wants to know if that is ok.

"Marge, would you please excuse me for a moment?"

She leaves the room.

"Rose, perhaps we should just tell your mother and the others that you are visiting with a sick friend from out of town? Izzy and I will be working night and day with the staff here to help all these people get their homes and lives back in order."

"I love you, Matt."

"I love you more than I can express, Rose, and pray that all this will be ok."

Rose is back in Florida. She disembarks from the plane and walks up the jetway through to the gate. She finds the nearest phone and puts in a few coins to call Paul.

"Hi Paul, who is picking me up here?"

"My son Danny will be down at the luggage carouse with your name on a paper. He has long blonde hair and looks like the cat dragged him in."

"Well, Paul, I heard Key West was rustic, but from the sounds of it things may be changing for us all soon."

"Rose, I don't want to say too much on the phone. The entire team will meet at the warehouse office by the marina this evening. Oh, by the way: Dr. Lawrence will be joining you on the ride out here. His flight landed already."

"Ok, Paul, I look forward to seeing everyone this evening. Take care and see you soon."

Rose then puts a few more coins in the machine to call Matt to let him know she has arrived safely and that Paul's son is driving her and Dr. Lawrence out to Key West from the Miami airport.

"Rose, by the way, your mom has called me twice today probing about your 'sick friend,' I think she knows something is up."

"Well, ok, Matt, don't worry; I will call her as soon as I get to Key West and settle in my hotel. You know her, she will keep me on the phone for an hour. I had better get down to the luggage area to meet Paul's son and Dr. Lawrence; he flew in today and is sharing the ride out to Key West, which I think will take about three to four hours. Matt, I still can't believe this is real."

Matt tells her to try to remain calm and focus on what needs to be done. He adds, "You know Paul, he can be hard to manage, and who knows what the rest of the crew is like."

"No problem, my love, I will be very careful with all of this."

They tell each other they love one another and end the call. She picks up her carry-on bag and heads down to the escalator and makes eye contact with a young man that looks like homeless surfer holding up a sign that says "Rose M."

"Oh boy," she utters to herself.

She approaches him and extends her hand to shake his and they both exchange an awkward hug.

"Well, Danny, I hear we have a lot of work to do."

"Rose, we cannot wait to show you what has been found, but the walls have ears everywhere and we need to be cautious about our discussions until we get into my truck."

"Have you seen Dr. Lawrence, Danny?"

"No, he came into international arrivals, so let's get your things and hop in my truck and head over to that terminal."

The two find her luggage and load the old Chevy station wagon. It is full of rust and has no air conditioning. The upholstery is worn out and torn and leaves a trail of smoke behind it from lack of maintenance.

"Rose, as you can tell from this car, things have been hard and lean for us down here. Wait until you see our salvage boat. That is even worse than this," he says laughingly.

"Wow, this is going to be some adventure," she replies, tongue in cheek.

Danny pulls up to the curb and asks Rose to watch the car while he heads in to get Dr. Lawrence. He had traveled light and has very little luggage.

Danny asks, "Where is your wife?"

"She will be here is a few days. She had come down

with a cold and wanted to get better before she flew and insisted that I get here as soon as I could."

"Thanks, Dr. Lawrence, we really need the whole team here right now."

As they head out to the car, Dr. Lawrence comments that he cannot believe the old junker made the trip in from Key West.

"With any luck we can make it back there today."

Rose sees them approaching and gets out of the car to greet Dr. Lawrence. She tells him she just cannot seem to get her head around all of this.

"How is Matt doing, Rose?"

"He is great but could not make the trip as his business is very busy right now, and as you can imagine he is worried sick about me."

"Well, all will be ok, I am sure, Rose. You have a lot of people to meet down there and they will treat us all like family."

"Yes, Key West is like no other place and once the word gets out about this things are going to get crazy in a hurry," Danny concurs. "In fact, I am sure we will have more 'friends' than we will need, of that I am sure. I have a cooler full of ice and soft drinks; sorry the car has no air conditioning."

The ride out to Key West is perhaps one of the most beautiful drives in America. There are more islands than you seem to be able to count connected by a series of bridges. The views are breathtaking. The color of the water is like no other Rose has seen before and ranges from emerald green to azure blue. The thought of treasure and the sites around were intoxicating to her senses. It did not matter how bad the transportation was; she was in a trance. Danny asked Dr. Lawrence to tell them about the book he had located that led them to change locations of the dive area.

"Danny, I was working very late and had many archives laid out on a table. I kept seeing that area mentioned over and over. What really had me convinced were the details written about the rescue and salvage attempt that had been made, which confirmed my thinking. It seems some of the crew of the Alucia had survived and made it to a small island in the area and were able to direct the salvage team but never were able to complete the job due to more storms that season, so all attempts to get anything from the wreck in those days were futile."

"Dr. Lawrence, what we have found is in relatively shallow water, only about thirty-five feet. There are ballast stones over a very large area, which makes us think that the ship ran over a reef in somewhat shallower water and ripped open a gap and started to spill the ballast stones as she broke apart. As she slowly broke apart in the storm we think the heavier load in the lower decks, such as the silver bars, all gave way at that time and went straight to the bottom in one area along the path with heavy metal objects like the cannons."

Rose hears the men speaking, but this all feels like a dream to her and their voices become muffled in the background along with the sound of the wind and an occasional seagull cackling. She misses her husband terribly and says to herself out loud, "If only my husband could see this view; he would give all the treasure in the world to be able to see what I can see now."

"Correction, my dear, I think he would give all the treasure in the world to see your face," Dr. Lawrence says.

Rose starts to cry and Dr. Lawrence says, "You know, Rose, I am sure he can see you better than anyone else on earth."

She smiles and wipes the tears away.

"Yes, Dr. Lawrence, I think you are right."

"Enough of this doctor stuff; we are all family now, please call me Martin."

"Ok, Martin, I am sorry for my emotional outburst, but I miss Matt and this is all very overwhelming to me."

The three stop on Marathon Key for a break and some gas. Rose walks to the water's edge and sits down on some rocks. She tilts her head back and closes her eyes and says, "God, please pinch me and tell me this is real and that everything is going to be all right."

The men come back to the car with some snacks and summon her back to the car. Once they are back and on their way she comments that the keys are somewhat of a spiritual place of sorts. It feels to her that she is on another planet. Danny tells her that the first time people experience the keys they are changed forever.

"I hope this treasure changes our lives for the best," Martin says. "History tells us that treasure has a way of conjuring up lots of wild, different things."

"Like what, for example?" Rose asks.

"Well, as soon as the word gets out I suspect poachers from all over the world will be scouring the area just like what happened during the gold rush in the American West."

"Ah, I see," Rose says.

"Danny, are you sure this word has not gotten out?"

"Yes, Martin, we have been very tight-lipped," Danny says. "Well, maybe. Uh-oh, man, look up there."

As the car enters Key West they can see police cars blocking the main bridge coming into the area. As their car makes it to a temporary checkpoint the police have set up, Danny asks the officer what is going on.

"Sorry, son, this is official business; we have to ask to check your license, registration, and insurance card please."

"Wow, must be a big drug bust in the area, huh?"

"I am sorry, son; as I said, this is official business. I will be back with your papers in a moment."

When he returns, the police officer asks, "How are you related to Paul O'Donnel, the registered owner of the car?"

"He is my father, sir."

"Son, is this the same Paul O'Donnell that runs the treasure hunting company?"

He clears his throat.

"Um, yes, sir; is everything all right?"

"Who are the folks with you, son?"

"This is Dr. Lawrence and Rose Messina; they work with my dad."

"Folks, I am going to have to ask you to come with me."

People are in the streets as far as the eye can see and it seems they are all heading toward the marina. There is a lot of commotion going on and they all demand to know what is going on.

"Please get in my car and I will explain everything," the officer says.

They get into the car and Martin says, "Well, the air conditioning feels good, but officer, what is this all about?"

"Well, it seems your company has located items of interest to the State of Florida. The state has seized your boat and is calling the attorney general."

"What! I demand our attorney," Rose says firmly.

"Calm down, please; he is at your office in the warehouse and we are going there now and will wait for the attorney general, who should be flying in within the hour."

"The state is claiming that all your findings belong to them." Martin says we are not saying anything more until we speak to our attorney: this is preposterous and an atrocity. The police car makes its way through the narrow streets of Key West. It seems like a festival of sorts. People were running around like it was the New Year's Eve or Fourth of July. They hear shouts of treasure and are running around like lunatics. The three are stunned and Martin motions to the other two by putting his index finger over his lips to indicate no more words are to be spoken.

As the police car pulls up to the warehouse they notice a barricade had been placed around it and a small army of police officers with rifles stood guard.

"Is this America or some communist state?" Martin asks.

"Sir, I am only doing my job," the officer says. "Please, let's all go inside."

Paul and the team, who appear to be in a state of shock, are all seated at a large table with maps laying on top. Shelton, the attorney, welcomes the three newcomers and says to the police, "We need our privacy. Can you all just leave and stand guard outside the door?"

The door closes and the group all starts shouting at each other; there is chaos in the room. Shelton bangs his fist on the table and says, "Please, everyone, we must be calm. We are not being held or charged."

"How did this news get out?" Martin asks.

"We are not sure, but I have some ideas," Paul says.

"My damn crew has a tendency to drink too much and I suppose maybe one of the divers flapped his gums in a bar, but no one has fessed up to anything."

"Let's keep our cool; we are not criminals," Shelton says. "The state is making a claim that they should be the owners of the treasure since it was found inside of Florida's waters. This may be a long legal battle, but let's just stay calm and wait for the attorney general."

The tension in the room was palpable. Paul is the most vocal and vitriolic with the situation, as he had the most time invested in this, not to mention his family, who had been in this long before any outside investors. Paul bangs his fist on the table.

"We have to root out the person who let this get out."

"Well, it was going to get out at some point," Martin says, "And we would have had to deal with this then."

"Yes, this is true, but we cannot have anyone on the team like this," Paul says.

Shelton asks Paul if anyone may have pocketed some of the treasure that could have gone unreported.

"No, I doubt it Shelton; I saw everything that came up. I guess some small items may have made their way into a diver's hands, but I really doubt it."

"Well, we have to make sure they all know that type of action would be considered criminal and would raise the specter of other things, as well as further erode our ability to stake a claim."

"I am sure they understand that," Paul says.

The police had been out rounding the crew up; they had all been getting bombed at the Sloop Wharf. They were placed in patrol cars and had now been brought back to the warehouse and into the room where the team was waiting.

"Hey, boss, what the heck is going on?"

"Well it seems that one of you idiots must have been bragging about the find and now the State of Florida wants all of it."

The men all look at each other and then back at Paul.

"Well, boss, we only told old Hammerhead, the bartender at the Sloop Wharf."

"You jackasses! What were you thinking? He is a bigger drunk than all of you."

"What have I gotten me and Matt into?" Rose mumbles to herself. "This is a nightmare."

"Ok, what is done is done," Martin says. "Let's pull

together. We must be tight from this moment on and listen to everything Shelton says. He is the boss now and Rose will have to detail every thing that comes up from the wreck, no matter how small or seemingly insignificant it may appear. I will work with Rose to detail the historical significance of each artifact. Do any of you have even the smallest of souvenirs that you may have kept?"

Two of the divers step forward and place a few coins on the desk.

"Did anyone see these?" Shelton asks.

"Yes, Hammerhead did."

Paul shouts out loud, "Lord Jesus H! No more of this! Do you guys understand?"

"Yes, of course, boss. We were caught up in the moment. It will never happen again, we promise."

With that, a sound of a helicopter is heard landing on the square outside the building. They all rush to look and immediately recognize the attorney general. He is led up into the building and into the room with the group. Shelton steps forward and announces that he is the attorney for the group.

"Sir, I have a cease and desist order from the governor. We have police boats which will go with anyone you select from your team to anchor and stand guard over the area until this is resolved."

"This is pure bullshit! Is this Nazi Germany?" one of the divers exclaims.

"Sir, I think it is best that you advise your client that they should exercise some restraint and composure. We are not here to take anything from your clients but rather we are doing our job to protect the interests of the people and State of Florida."

"Sir, we understand you have a job to do, and in the spirit of full disclosure I must tell you that I am an investor as well and that we will likely be bringing in additional legal support."

"Thank you for sharing that, sir; here is the stop order and the state police onsite will accompany your team to the wreck site to protect it. I will be in touch in the morning to get things moving quickly."

"This is going to be a media nightmare and put a lot of pressure on this tiny community with the onslaught that is about to arrive here," Paul says.

"Well, there is a silver lining to all of this," Rose says. "If we need more investment capital to keep this expedition going, the media coverage will bring in lots of new money."

Chapter 9:
FLIGHT

Back on his personal jet, Matt is awakened by some turbulence and calls for Michelle.

"Do me a favor; ask Blake to come back here and join us, would you please?"

"Yes, of course, Mr. Messina, and she dutifully heads to the flight deck and tells Blake the boss would like to see him back in the cabin.

"Sure, I will be right there."

As he takes his headset off he tells his copilot to take it for now.

"I will see what this is about."

Blake joins the men and Matt asks him to sit down and begins to tell him of his day and that he had a chance to captain his yacht today.

"Blake, it would be a huge honor if you let me have the left seat for a bit and you sit in the first officer chair.

You know, Blake. I love to fly, and I think I was once told that you guys have to learn to fly 'blind,' whatever that entails. Is that true, Blake?"

"Ah . . . yes sir." Micheal makes eye contact with everyone.

"Good, so let's do this! I didn't want to have this discussion up in the flight deck with your first officer; I don't know him as well as you, Blake."

"Well, sir, I have known you for a long time, so let me deal with some things up front and I will page Michelle when I am ready for you."

Blake heads back to the cockpit and tells his copilot of this.

"You are not to breathe a word of this to anyone. Mr. Messina is our boss and meal ticket and I am your boss. I want you to go into the rear galley so you have no witness of this. I will page you when I am ready for you."

"Blake, this is nuts, but whatever you say."

The first officer walks through the cabin and wishes the men a good evening, to which they respond, "And you as well," along with some light laughter. Michelle joins the first officer in the rear galley and closes the door.

"Hey, pervert, don't you even think about it; we are just giving them some privacy up there."

"Sure, honey, whatever."

A few minutes pass and Matt pages Michelle on the intercom.

"Ok, bring the boss up."

She tells the first officer, "Please stay back here. Mr. Messina is serious about this."

She heads into the main cabin and says, "Sir, Blake is ready for you up front."

Blake greets him at the flight deck door and says, "Sir, we are in auto pilot for the moment. I want to acquaint you with your surroundings here, as you will need to be a bit of a gymnast to get situated into your seat. Please, if you don't mind, let me take your right hand so that I may raise it to give you an idea of the space and headroom around you."

"Sure, Blake."

With that he raises Matt's hand and says, "As you can tell, we have very limited space in here. The roofli-ne slants forward and is loaded with switches."

Matt gently passes his hand over the switches.

"I understand what you mean, Blake."

"Ok, sir, now directly in front of you I will guide your hand to the controls which are between the pilot and copilot seats. You will need to clear this cluster of instrumentation when getting into your seat."

"No problem, Blake, I am more limber than you may think."

"Ok, to your left is your seat; please put your left hand on it to position yourself."

"Ah, no problem, Blake."

"Ok, sir, one last thing before you get into the seat. Reach forward to feel the yoke; this is sort of a steering wheel. It goes to the floor on a post that your legs will straddle and on the floor are pedals to the forward position; please don't put your feet there, just keep them resting on the floor. So, sir, you all set?"

"More than you know, Blake."

"All right. Michelle, I am going to get situated in the right seat and maintain control; please help Mr. Messina should he need assistance into his seat."

"No worries, Michelle, I think I can do this; just stay there if I need you."

Matt raises his left leg then his right and slides into the seat like a pro and buckles his seatbelt.

"Ok, sir to your left is a headset. You will notice a microphone sticking out on the side; that belongs on your right."

Without hesitation Matt reaches and places it on his head. Michelle and Blake exchange smiles and Michelle says, "Ok, I think you men are in charge here; I will leave you two and head back to take care of the others."

The cockpit door closes and Matt turns to Blake, "Hey, do I detect that the two of you are an item?"

"Well, sir, not yet, but I am trying."

"Good luck, I wish you the best."

"Thank you, sir. Now let's get to the business end of this plane. There are a few basic things I want you to know. First, the yoke does a few things, including the climb and descent of the plane. Pushing it forward will make us lose altitude and pulling it back will make us climb. Turning it to right will tilt the wings to the right and the same for the left. The pedals on the floor make us turns to the right and left. I will explain that a bit better. Give me your right hand. The pedals control yaw, and that's what this feels like."

He mimics the motion with Matt's hand, holding it flat and level, and moves it to the right and left.

"Ok, now, the wheel makes the wings tilt."

Blake tilts his hands to the left and right with a tilt.

"Ah, I feel it, Blake."

"Ok, so to your right are levers that you stepped over to climb into your seat; those are the speed controls—similar, I suspect, to what is on your yacht."

He places Matt's hand there.

"Yes, I think they are similar."

"Ok, so there is a whole bunch more, of course, but this is what you will need to understand to get a feel for this."

Blake gets on the intercom and tells the others that they will be making some gentle maneuvers and to please stay seated and belted.

"So, sir, please just gently place your hands on the wheel and your feet on the pedals—but very gently, please. I just want you to feel what I am doing. Everything here is duplicated, so controls are on both sides, but we share the speed controls in the middle."

Matt complies and places both hands on the wheel and feet on the pedals.

"Ok, sir, I am going to turn off the auto pilot and make the plane tilt to the right and then to the left. You will also feel me pull back slightly, as planes tend to lose altitude when you tilt and turn; this will compensate for that. Ok, so here we go."

Matt's smile is ear to ear as the plane slowly turns to the right and left.

"Did you feel that, sir?"

"Yes I did, Blake. Can you do that again, please?" The process is repeated and Matt says, "Ok, may I give it a shot?"

"Of course, sir, I am right here and will take control if need be."

Matt clears his throat and Blake interjects and says, "Sir, it's important that less is more; make very gentle movements. An airplane is like a woman, if you get my drift."

"Got it, Blake; here we go."

Matt turns to the right and Blake reminds him to pull slightly back on the yoke as to not lose altitude, recalling that a plane tends to do that in a turn and pulling back on the yoke corrects this.

"That's perfect, sir, now let's try that to the left, but first I want to keep us on our heading."

Blake gets them level and on course, after which Matt repeats the process of left to right.

"Ok, sir, now let's get the feel for ascending and

descending. Ok then, sir, I am going to make us climb first. Keep your hands on the yoke and feel me ever so slowly pull back on it. Right now we are at thirty-two thousand feet and I am going to bring us to thirty-seven thousand, but give me a second. I want to get some clearance from air traffic control. The maneuvers we did earlier did not require clearance, but this will in this space."

Blake contacts air traffic control and requests clearance for the next ten minutes; he says he would like to fluctuate altitude between thirty-two thousand and thirty-seven thousand feet to test some performance standards of the plane. Clearance is granted and Blake slowly climbs to thirty-seven thousand feet. Matt comments on how subtle the motions are to get the plane to react.

"Oh, yes sir, for sure."

"Blake, it seems to me that with all the computers and support from air traffic control you literally could fly blind."

"Sir, for much of this you are somewhat correct. When we get back from Italy let's spend some time in a small plane, ok, Blake?"

"Yes, I would like that, sir."

"I think we should drop the formalities; call me captain, since I am in this seat."

The men both laugh.

"But really, Blake, call me Matt."

"Ok, Matt, now you are going to take us back down to thirty-two thousand feet and I will watch our heading."

Matt gently pushes forward on the yoke while Blake calls out the change in altitude every five hundred feet to give matt the sense of where they are.

"Ok, we are back to thirty-two thousand feet, Matt. The only thing we did not try with the basics were the throttles. Pushing them forward will increase our speed, and pushing them back will slow us. So, Matt , my left hand is on the throttles. Reach out with your right and put in on my left hand. Ok, now I am going to take us from four hundred twenty-five miles per hour to four hundred fifty."

Blake slowly pushes forward and hits four hundred fifty with little effort.

"Ok, Matt, now pull us back to four hundred twenty-five and we will keep our hands as is."

Matt pulls back and once they hit four hundred twenty-five they repeat the process. Matt raises his left hand and places it on the window to his left, sensing the wind rush by.

"Takeoff must really get you guys charged up."

"For sure, Matt, it does; it never gets old. I have to let the air traffic control folks know our maneuvers are complete for now."

He communicates the same over the radio.

"Hey, Blake, you mind if I just hang out here for a bit more? I am really enjoying being up here."

"Matt, I am having fun, too, and after all, this is your aircraft."

"Well, Blake, maybe someday you can have one of these. Maybe we should partner on a charter of some kind. You bring the eyes and experience and I will bring the cash."

"Matt, are you serious?"

"Oh yeah, Blake, as serious as a blind guy flying a jet."

They laugh a bit and Matt says, "Hey, we both love to fly and I know a bit about business. I have been around you over the years and know you have the passion and dedication to guide us on the aviation side of things. So that's it, let's give it some serious consideration, ok, Blake?"

"You bet, Matt; I sure will."

"All right, then, let's chat soon after we get back."

"You better call for Michelle to take me back to the others. They must be white knuckling their seats back there."

With that Blake calls for her and tells the others that they may relax as the flying lesson is over. Michelle opens the flight deck door.

"Well, how did it go up here, gentleman?"

"Well, you may just want to gauge that by the smile on my face. How are my boys back there?"

"Sir, they are fine, but, frankly, I think they are happy the lesson is over."

"Well, if they want to fly back with us they will have to deal with my next lesson, which I heard is a bit more involved. In fact, I heard pilots have to learn to fly inverted."

They all get a good laugh from that remark.

"Ok, Michelle, take me back to my crew back there. Blake, thanks very much; I am very grateful for this."

"Matt, it was my pleasure; Michelle please have our buddy come back to the flight deck to take the right seat; we are about an hour from our final destination and I would like him up here to prepare for our approach and arrival. See you on the ground, Matt."

"Thanks again, Blake."

Matt joins the boys, who are looking through the family album, and Mark greets Matt with a, "Well if it isn't Charles Messina Lindbergh. That must have been some experience, dad. I can assure you it was for us. What else do you have up your sleeve: hot air ballooning over the Sicilian countryside, perhaps?"

"Hey, that's a great idea, son; I hadn't thought of that."

Luke is quick to interject: "I think it's time for some grounded fun, guys."

"I was kidding, Luke . . . but you never know."

"Dad, Mark and I have been looking though this album and we are excited about the trip. We had forgotten how beautiful it is there. We had so many great times there. We really should go back more frequently."

As he is saying this Mark is looking down at a picture of his mother holding freshly picked flowers. She loved flowers; they were one of her pastimes. His hand gently touches the picture of her face.

"Mom looks so beautiful in these pictures, dad; we had so many nice times as a family there."

"Yes, son, I am very sure we need this trip for more reasons than we can imagine. Get ready for an onslaught of food and people that love to be with family. Giovanni, your cousin, will be picking us up at the airport; do you guys remember him?"

Luke is quick to recall some of Giovanni's antics.

"Oh yeah. He is quite the prankster. He would tease all the young girls and would do anything for a practical joke. I wonder how he and Rose are getting along. She used to be the target of many of his little schemes. Do you remember when he 'accidentally' spilled a bowl of pasta on her at the kitchen table?"

"Well, son, he is much older now; let's hope he has matured a bit. I heard he is in law school now."

"I heard something about that," Mark says. "Maybe he is learning how to get himself out of jams."

The three men get a good laugh at that. Blake announces that they will be descending soon and they should buckle up because the ride in will be a bit bumpy the rest of the way. Michelle comes out to clear all their glasses and loose items and asks if there is anything they need before they settle down for the landing.

"Yes, just please tell us that our father is not going to be allowed to land this thing."

"You know, you all keep giving me great ideas."

Sicily is soon in view out the windows of the cabin.

The boys remark how great it looks and how beautiful the water there is. Luke comments that it's almost like the color of the water in the Keys.

"Yes, your mom made the same observation, son. We have to go back down there soon to visit with our friends. I heard Paul is not well and is battling illness. We really should see what we can do to help him. His family is really struggling with his illness. This is a good example of how you can have all the money in the world and it still cannot buy health or happiness. I spoke to Paul recently and he told me he is at peace and the legacy of the Alucia will carry on."

Blake comes back on the intercom to let them know that their landing will be delayed slightly due to some traffic issues at the airport but they have priority due to their fuel situation.

"Well, dad, did you use up all our fuel with your antics?"

"Ha, no, I am sure we have plenty; Blake is the safest pilot you will find. While we have the extra time up here, boys, lets go over a few things. Rose is a very sensitive young lady and I want her to enjoy her time in Italy. I don't want her worrying about me or things back home. So, remember, I am simply going to tell her that Dawn is visiting with a sick friend from college. I suppose this way she will not have anything to relate to about the person that Dawn is 'visiting'."

Luke and Mark agree that is good idea and they will stick to that story.

"You boys should think about seeing the country while you are here. The culture and food are amazing. I would think Rose would be thrilled if you hung around for a while. I can only stay for a week or so. There is such much happening with the business and I will be tending to this situation with Dawn. Rose will become inquisitive if I stay too long without Dawn being here. The longer I stay the more she will ask; she is very intuitive."

Mark tells his Father he should let others handle that and he should be with family.

"Son, I understand your thoughts, but there are certain things other people can do and some I must follow through on myself. Let's relax and enjoy this coming week."

Matt asks if there is a picture in the album of the family on the steps of a church. Luke flips though some pages as he says, "Yes, I think I saw that in here. Ah, yes, here it is."

"Boys, that is the place your grandparents were

baptized, confirmed, and married. That church is very old and your bloodline helped to build that. We will be going to Mass there this Sunday. The best part of that is the food after. It seems like the entire village pulls together and there is one huge food fest. You boys were too young then but now you will be able to enjoy the best wine in the world. In my opinion the Italian reds are among the best on the planet and the people who make them put a lot of love into it. Family, food, and wine: the staples of Sicily."

Luke comments that it all sounds great, but his eyes and palate will be seeking other things while he is there.

"Oh yes, this place is magical for that. Just make sure you boys keep it out of the family. I bet the vast majority in our old village is related in some way," Matt says with a laugh. Blake announces they have been cleared for landing and to strap in tight as they expect some high cross winds as they come in. The plane is soon experiencing severe turbulence.

"Wow," comments Mark, "I am not sure what is worse: this or experiencing the hand of lucky Lindberg's flying lesson."

"Ah, come on boys, this is nothing."

Things begin to rattle and fall out of cabinets. Michelle lets out an audible gasp from her jump seat in the galley. The sound of the landing gear is heard kicking in with the sound of wind racing past the extended flaps. The plane touches down hard and the reverse thrusters come on almost immediately. They are down safe and Blake comes on to announce," Welcome to Palermo, folks; I will have us into their assigned area in a couple of minutes. It seems the controllers here are still dealing with a lot of ground traffic."

"Hah, the controllers are probably having a nap; they have too much wine in the tower, not to mention that Italians have a way of making their own work hours. You boys will see that very soon," Matt says.

Blake comes on again to apologize for the rough landing, but states it was best to get the plane on the ground rather than divert to another location given their fuel situation. The plane comes to a full stop and the engines are shut down. Michelle comes over to assist the men with their things and Blake comes back to wish them well.

"Hey, if you are not busy, why don't you and the crew join us for Sunday dinner?" Matt asks. "It starts at three and ends when we can't eat or drink anymore. There is always more than enough to go around."

"Matt, that would be a great honor and I accept the invitation on behalf of the crew."

The first officer takes notice of Blake calling the boss by his first name.

"Wow, you two bonded up there."

As the family deplanes, Blake comments to the other crew members that he has been flying private jets for almost ten years and no one has ever treated him as well as this. He states that rarely does any private passenger treat him like anything more than a chauffeur. He tells his crew that they should bring something with them to the dinner.

"Michelle, you were close to their conversations; did you hear anything they like?"

"Yes, Mr. Messina is fond of Italian Red Wines, I think Amarone specifically.

"Ok, lets make sure we take the Messinas up on their offer and show the respect of bringing what they like."

Chapter 10:

The Old Country

Once on the tarmac Mark spots Gioavanni waiting behind the fence, leaning against his car. He is a very handsome young man. As the men approach, Giovanni throws his cigarette to the ground and walks quickly to greet them. He hugs Matt and kisses both his cheeks and says, "Dear uncle, it has been too long! We all miss you and the family."

The boys exchange handshakes with the dashing young Italian.

"Cousin, it's great to see you," Luke says. "How is our sister? She loves it here and is being treated like a queen."

"Uncle, you would be very proud of her. She conducts herself with a humble confidence. It seems like all the young men in our village would like to court her and she has not allowed one to date her. She is very focused on her studies and her future in your family business."

"That's our Rose, very much like her mother," says Matt.

"Uncle, where is Dawn?"

"She will not be joining us on this trip as she has been called to help an old college friend who has taken ill back in the States."

"Ah, I understand and hope that her friend gets well."

"Thank you, Giovanni; is that cigarette I smell? You should quit that, young man."

"I know, uncle, it is a nasty habit and one that I have tried to shake."

"You keep at it, young man. Rose has no idea we are coming, right?"

"Yes, that is correct; only my father and mother know, and by the way, they do not want you three to stay in a hotel. We have a guest suite that is not being used."

"That is very kind of the family, Giovanni; we would love to be closer to everyone while we are here. Before I say yes, what do you boys think? I would think you boys may have designs of your own."

Mark and Luke agree it would be great to have

everyone together in one place as long as Giovanni will not be pulling pranks on everyone all the time.

"Cousin, I am over that, but I will tell you I do like a good time and we could spend some time downtown in my favorite places."

They pack the luggage into Giovanni's car and head home.

"Well, guys, it must be really something to fly in a machine like that. Maybe you can teach me about business, uncle? I am studying at the university and my primary area of focus is economics."

"That is very good, Giovanni. I insisted that my children all go to college and furthermore I also have a very strict policy in the business. Before anyone in the family can come to work in the family business, they must first work at least three years somewhere else. It is important to learn how other companies function and how to work for good and bad bosses. There are many great things you will learn in college but certain things cannot be taught in a classroom environment. There is another thing that is required in our family business: once a family member joins the firm they must report to a boss that has the ability to fire them without reprisal from me. Nepotism is like a poison to business because it incites bad morale and mistrust within the ranks. We also require thirty hours of community service each year from all of our employees. They may select from a number of philanthropic organizations and they are allowed to take time from work to do this. It is important to give back to society; in fact our family started a foundation to help the disadvantaged masses of Haiti. Many businesses lose sight of what is most important: service to not only our customers but our fellow man. Many companies will send a check for a tax deduction and say they are doing good and that is fine but I suggest it is far more important to have people being engaged in service."

"Wow, uncle, that is very interesting. I have learned things already."

"My dear Giovanni, we never stop learning. In fact, I have learned many things lately that I trust will sharpen my focus on certain things. In fact, that is another reason we are here. The basis for a successful enterprise starts with having a solid foundation at home. Leaders of business must have a stable supportive core at home if they have families. We must never let so much time go between our visits with each other. At the end of the day, what is most important is family and no one knows that better than the Italians. In the States it seems that things are faster and more fragmented than they are here when it comes to family. I love America and all the opportunity that it has given us. I am very proud to be an American. That is another thing my father taught us in America. We remain very proud of our Italian heritage, but my Father would discipline us if we spoke Italian in public or did things like putting Viva Italia bumper stickers on our cars. One of our toughest times was when some of our relatives had to come back to Italy during World War II and oppose our own. My point of all this, Giovanni, is that we are grateful to be here; let's do all we can to remain connected as much as possible."

"Yes, uncle, we must do that."

"By the way, Giovanni, this car stinks of cigarette smoke."

"Yes, uncle, I will be sure to clean it up."

He smiles at his cousins and they return the smile as the car bounces down the dusty road leaving a dust plume behind it.

The car is bumping around a bit and Matt starts to laugh with a sinister tone.

What's so funny, uncle?"

"Hey, Giovanni, you have been known to be a

joker and I understand that you just told us that you are over it. Well, I am not sure of that: once a joker always a joker."

" Well, uncle, that may be true but what I meant was that I no longer waste time with childish pranks."

"I have a great idea, nephew."

"Oh boy, here he goes again; what now, dad?" says Luke.

Matt asks Giovanni if Alfredo's place is still open down the road from the village.

"Yes, uncle, he is still there and he is growing old."

"Ok, it's settled; lets stop for a some olives, cheese, and wine."

"Uncle, everyone is waiting for us."

"That's ok, we will not be long. I want my boys to experience Alfredo's. It is such a great example of what this area is all about, but more importantly will give us a chance to discuss what I would like to do as we arrive in the village."

"I can't wait to hear this one," Mark says. "Dad has been on a bit of a roll the last couple days." The car makes a right turn up a steep hill where the restaurant overlooks a beautiful valley. Giovanni's dashboard plastic Jesus statue falls down from the incline, to which Luke says, "Oh, that is not a good sign."

"Oh, no worry, cousin, my Lord and I have been through far worse."

Before they get out of the car Matt asks the boys to pause and smell the olive grove.

"Speaking of the Lord, he wants us always to take the time to enjoy these things."

"I must say, it's better than the smell of your car, Giovanni," remarks Luke.

They all laugh and walk into the old stone structure; as soon as the door is open the wonderful scent of fresh tomato gravy and baked bread fills the air. An old man calls out from the kitchen, "Someone will be right with you."

"We're in no hurry," replies Matt.

"Oh my, is that Mateo Messina?" calls out the old man. He rushes out of the kitchen with his stained apron and hugs Matt and then kisses him on each cheek.

"My friend, you are a sight for sore eyes! Oh please, excuse me, Matteo, I meant—"

"It's ok, Alfredo, you are a sight for sore eyes, too."

The men all laugh.

"Matteo, I know Giovanni, of course, but are these handsome men your boys?"

"Yes, Alfredo, fortunately they took after their mother in the looks department."

"Hello, sir, I am Mark and this is my brother Luke."

They exchange handshakes and hugs. Matt says, "Before we get settled and I forget, Alfredo, please put together a few nice baskets of bread, cheese, and olives. I want to make sure I arrive in town with some nice things for the family.

Alfredo says of course and mentions that he heard Matt's daughter was already in town.

"Yes, she is, and we are coming to surprise her and reconnect with family. Only a few people know we are coming."

Alfredo is quick to comment that he would not be surprised if more people knew about their arrival, as the family would be too excited and would prepare for such things.

"Well, we hope not, my good Alfredo; please join us in a toast to old friends and our health."

"I will, but first let me go back and fix some special snacks and I will treat you to my best bottle of wine."

"I knew this was going to be a great stop for us, men," comments Matt. "I also want to tell you of my idea about our arrival in the village."

"Well, uncle, before you do that maybe I should tell you that Alfredo is right. My parents told Rose a little white lie and said that relatives that she had never met were visiting from Sardinia."

"Why would they do that, nephew?"

"Well, uncle, they have prepared a welcome sign for the town square and have been cooking for days. Dozens of friends and family will be there to greet you and the boys but Rose has no idea it is you that is coming."

Alfredo walks up at this moment and places the food and wine on the table. He has overheard what is happening and says, "You see, Mateo; I am not surprised. You and the boys are in for a real treat." As he pours the wine he tells them that the cheeses had all been made in the region and that the olives were hand-picked on his property by his family. Matt stands up with his glass and the others rise to toast.

"Well, men, this is a special moment, and while I expected a warm welcome I did not anticipate what we are about drive into. I have an idea that will make our arrival even more fun. It is time this blind man drive a car."

"Here we go; I knew he had something up his sleeve," Luke says. "You see, gents, in the last forty-eight hours dad has driven a boat and flown a plane, so it only stands to reason that he wants to drive a car. So, dad, let's do it!"

"Cheers to family and fun!" Mark toasts.

They connect their glasses and enjoy the finest wine in the region.

"Ok, dad, tell us of your idea on driving a car."

"Well, I have a story to tell you men. After a few of my buddies returned home after World War II they came over to pick me up to drive through town to look for some fun. Well, one of them came up with the idea that maybe it would be hysterical if I had driven the car. So one of them squeezed down on the floor and operated the gas pedal and brakes while another sat close and helped me steer by placing his hand on the bottom of the steering wheel so no one could see it. I had my left arm on the side of the door with the window rolled down. We drove slowly around the streets of New Brunswick beeping the horn at people we knew and greeting them with shouts that the war was over. The guys with me in the car said people looked like they had seen a ghost or something. My buddies could not stop laughing for days."

"Ha, you see, uncle, now I understand where I get the prankster from."

Alfredo insists that he, too, must join in the fun and walks back to the kitchen to tell his staff that he will be gone for the rest of the day. The men spend a few more minutes enjoying the food and discuss the arrival plan.

Giovanni suggests that they pull off the road just before the entry into the town's main gate to allow for positioning themselves in the car just as Matt's friends had done back in the States. Alfredo asks for a few more minutes to get the baskets ready.

"Ok then, boys, let's go outside and rehearse this a bit."

Some patrons arriving at the restaurant see this taking place out in the parking area and one comments, "I heard the wine is good here but that is a bit much." The car is making slow right and left turns in the parking lot, reminiscent of what Matt had done in the plane.

Luke looks at Mark and says, "Does this remind you of something, Mark?" More onlookers see this as they enter the lot and one recognizes Alfredo as he brings the baskets to the car.

"Hey Alfredo, what is going on here?"

"Well, it seems we have a blind man learning to drive."

The folks who had asked look puzzled as the car comes to a halt.

"Looks like this is going to be a bit cramped in here. You fellows put the baskets in the car and I will follow you into the village on my scooter."

What a sight! Alfredo must be in his seventies now and is riding behind the car on his scooter with his dirty apron flapping in the wind. The dust is kicking up behind the car, bathing him in a sea of dirt, so he decides to pass them. As he does Giovanni shouts to pull over into Faracabalo's at the bottom of the hill. The small parade makes its way down the hill where the views are breathtaking. Matt's sons are taking this all in and the buzz of the wine has put everyone into a happy mood. Faracabalo's place comes into view. It is a vast estate with vineyards on each side of the long driveway. Alfredo pulls into the gate and pauses as the car pulls in beside him. The men take their places in the car to stage the blind driver's arrival. Matt is behind the steering wheel. Alfredo is lying across the front seat with his torso draped over onto the floor and his arms extended to help Matt steer with one arm; his other arm is near the gas and brake pedals should Matt lose control. The boys are in the back telling Matt where to steer. They all decide this is the best format and most likely to fool people as no one else will be visible in the front seat. The drive into the gate of the village is less than a mile from them and only a slight bend in the road needs to be navigated. They pull out onto the road very slowly and this time Alfredo is alongside the car so Matt can hear his position as a reference point, much like a wingman would in a plane. This scene is hilarious as the formation comes around the bend and the gate is in view. A sign that reads, "Welcome Bongiovannis," is seen hanging there. As they slowly enter the gate a group of people can be seen in the town square and music begins to play from a small band. Rose is there alongside of her cousins and cannot believe her eyes. She looks at her aunt and uncle, who now see that she is surprised. Matt's left arm is hanging out of the window and the other arm is "steering" the car as it comes to a stop.

Rose shouts, "Father! What are you doing?"

"Bongiorno, my love!" The band closes in around the car and Rose also sees the boys at this point. All get out and are hugging and kissing. The relatives are crying at the sight of the family reunion. As this is taking place Alfredo retrieves the baskets from the car and starts to hand out wine and cheese to all that have gathered.

"Leave it to my Father to make an entrance like that!" Rose says.

"That is nothing; wait until you hear about the pilot and boat captain that he has been of late," Luke says.

"Father, have you been playing a bit?" asks Rose.

"Well, honey, I am just catching up on some things I wanted to do."

"Hey, where is Dawn, guys?"

Matt says she had to visit with a friend from college who had taken ill.

"Honey, we wanted to surprise you and spend time here with the family. We just came up with the idea and flew over here."

"Wow, that is so great, dad, they really had me fooled!" I though some long-lost relatives were coming in from Sardinia."

The crowd begins to get larger as the word spread that the Messinas were back in town; many more wanted to join in the festivities. Luke and Mark are quick to notice several young beautiful ladies; winks and gazes are exchanged. Mark yells over to Giovanni: "You are going to have to help us learn which of these fine-looking women are available to date."

"You boys don't waste much time; relax and enjoy. They aren't going anywhere."

"Yes, I guess you are right, and we have had some wine."

The band switches to dance music and soon everyone is locked at the arms in a big circle; they all begin to dance. The first dance exhausts Matt. The last few days have been heavy on his heart and hard on his body. Rose notices his fatigue and takes her father by the hand and walks him over to a bench next to the fountain.

"Dad, this is the greatest surprise of my life. I am so happy we are all here together." Tongue in cheek, she adds, "Well, um, I am sorry Dawn couldn't make it, dad."

"Honey, it's ok; you know, this may be a good time for us to bond and remember some of the special times we had here with mom."

Rose is a bit curious about Matt's reply and looks at his face while taking his hand and kissing him on the cheek.

"Yes, dad, this will be a great time. Hey, I see Alfredo followed you in. I guess you stopped at his place on the way in?"

"Yes, we did; I wanted the boys to see his place and to take a break to plan our little arrival."

"Dad, you are so funny. We are so lucky to have you as our father, and I hear you have been working really hard at the office. I think it is time you took a break and I am glad you decided to come. How long do you plan to be here?"

"Well, honey, there are some things that need my attention back home and I was thinking about a week, but the boys can stay longer."

"Dad, I insist that you consider staying longer. I haven't been outside of Sicily since I have been here. Maybe we can head off to Cinque Terra, Tuscany, and some other areas that mom and you loved?"

"Well, honey—"

"Don't say no so fast; we can talk about this later."

Rose explains to her father that the Messinas have become famous in the little town as they are the first descendants to have become extremely rich, and while money did impress the villagers it was not more important than the love of family and friends. She goes on to say that the local wisdom knows that money comes and goes but family is the bedrock of one's existence. Money only makes things easier.

"Honey, this is one of the reasons I wanted you to come here and learn these lessons. I want the boys to learn and feel this, too. Back in the States it is so easy for us to lose perspective of what are the most important things in life. Your mother and I would preach this often and I am sure you understood what we were saying but being here reinforces it in so many ways that we could not do at home. Maybe you are right, I should consider staying here longer. Your mother did such a great job of keeping everyone together and we are exposed to a lifestyle that tends to distract us from our core values back in the States."

The boys notice Rose and Matt sitting on the bench and come over to join them. With this the villagers who are not part of the family come over to introduce themselves, some as families and couples, and others as individuals. Each has something for them. Some have made little trinkets and others have made gifts of baked goods and fresh food. The Messinas are completely taken back by this. Their emotions range from humility and happiness to pure joy with the welcome. The last to acknowledge them is young boy of about nine years.

He walks up to Matt and for a moment stares at him, noticing his blindness. The boy does not say a word but reaches up to take Matt's hand and holds it for a moment. Rose and her brothers look on, as do the others still in the immediate area. Matt puts his free hand on the boy's hand that is holding his and says, "Hello, young man," in Italian. The boy is even more curious. He asks Matt, "How do you know I am a boy, sir?" Matt says his Italian is a bit rough and Rose says, "Dad, I will tell him. What do you want to say to him?"

"Explain to him that I have two boys and a girl; they are seated here to my right. Explain to him that I used to hold their hands when they were young and a father can tell the difference between the hand of a boy and a girl."

Rose translates this for the young boy, who then looks at each of the Messinas seated there on the bench. Matt asks him his name, to which he replies, "Salvatore." Matt tells him that is a very special name in his family, as his father was named Salvatore. The boy still holding his hand says in Italian, "I think you are able to see things that other people cannot, sir." The family is stunned by the boy's intuition. Matt is able to understand what the boy has said.

"Thank you, young Salvatore; perhaps you are right."

"Sir, I am so sorry I do not have a gift for you and your family."

"Rose, please tell him that he has brought us a gift, more than he knows. Tell him that I would like he and his family to join us at Sunday dinner this week."

This is offered to Salvatore, to which he requests whether he may bring his mama, as that is the only family he has. Matt understands what the boy has just said and stands up then kneels beside the boy and gives him a hug and is able to put the Italian words together to say, "Please, bring your mother."

Salvatore thanks Matt and walks away. Matt stands up as if to stare at the boy as he walks away. The moment is powerful and emotional for all the Messinas.

"Dad, I am so glad you brought us here," Mark says.

"Me too, Dad, me too," Luke agrees.

The sun is beginning to go down and the party music is winding down. It is time for rest after a long journey. Alfredo comes over to tell them that if there is anything they need while they are there to call on him. Matt says, "Please have your family come join ours for Sunday dinner."

"Yes, Matteo, of course we will be there. I know

your hosts well and will help them with the preparations; from the look of things we will be having many."

"Yes, Alfredo, I even invited my flight crew as they have never experienced a true Italian feast with food and family."

"I am sure it will be a special day, Matteo. You all look like you can use some rest. I will leave you now; please enjoy your time together."

"God bless you, Alfredo; safe trip back up the hill, my friend."

As the sun sets over the village the band keeps playing. Many linger around the fountain but the Messinas are all exhausted from the trip and the welcoming party. Their hosts show them to their rooms, where they retire for the evening. After they are all settled there is a knock on Matt's door. It is Rose to thank her father again for the surprise visit.

"Rose, I feel your mother's presence here. I haven't felt that in a long time. I have never stopped loving your mother; I hope you know that."

"I know, dad, I just hope you are happy in every way."

"I am, Rose, especially now that I am here with you and the family."

Rose kisses her father and wishes him a good night. She does recognize he did not mention missing Dawn but lets it go for the moment. Matt lays his head on the pillow and realizes Rose may know something is wrong. The two are very connected on many levels and not able to hide things from each other. Matt gently strokes his left hand over the sheet while he imagines his dear wife beside him and begins to weep as he drifts off to sleep.

Meanwhile, back in New York City, Matt's legal staff is in high gear, seated in the conference room as they wait for Jeff to enter the room.

"May I offer you some coffee, Damiano?" asks an aid.

"If you don't mind, may I have a diet soda?"

"Yes, of course, any kind?"

"No, whatever you have that has caffeine is fine, thanks."

With that Jeff enters the room with his note pad and sits across the table from Damiano.

"So, Damiano, did I hear the Messinas are in Sicily?"

"Yes, Jeff, they decided to get away from this and join Rose as she has been there for an extended time to study and learn the culture."

"So, I take it the boys are there as well?"

"Yes, the last few days have been quite the whirlwind."

"Damiano, I will need to have direct conversations with Matt as soon as possible. I have read the notes and have requested that the family trust files be brought to me along with the prenuptial agreement for review. This all seems rather fast, though, and such aggressive action is being taken, I must ask: is there no consideration for marriage therapy or a cooling-off period? Damiano stands and looks at everyone in the room without saying a word and then sits back down. Jeff knows that when Matt has spoken the decision has been made and a plan of action must be executed. Damiano adds: "I will brief Matt's brother on every detail of this matter. He is not to be disturbed as he is focused on closing a significant deal in Dubai."

"Damiano, I would like you to meet my most trusted investigators. This is James Gordon and Dean Westley. They are without any doubt the best this practice has at its disposal. They will need to know what you know and we would like to have a conference call with Matt as soon as is practical. In the meantime, we will get our hands on the surveillance records from the locations that Dawn frequented that had such devices."

Damiano reports that he suspected that Shamus was likely to have deactivated it at the family residence on Park Avenue. He says it seems that Shamus was covering for Dawn.

"For those of you who don't know him, he is a down-on-his-luck kind of guy who is the brother of Dawn; Matt employed him as doorman at the Park Avenue home. You see, Shamus is not the brightest bulb on the planet and has not been able to hold any jobs. Matt offered to help him and for years he was great, but I would think Dawn had talked him into covering for her antics."

Jeff orders Dean and James to go keep an eye on him as well.

"Damiano, do we have any idea who Dawn's lover is at this point?"

"No, Jeff, and Matt doesn't either; but he also made it clear that he is absolutely done with her and has mentioned that he was giving thought as to a plan of action for dealing with her. All the locks have been changed on all the residences except for the one in Jersey City, New Jersey."

Jeff asks his assistant to get Matt's private bankers lined up for a meeting within twenty-four hours, and as he is saying this a secretary enters the room and says Matt is on the line. Damiano looks at his watch; it must be about 10 p.m. in Sicily. Jeff picks up the phone in the conference room, where a rather tired-sounding Matt is on the other end of the line.

"Hello, Matt. Damiano and my team are in the conference room. How are you, my friend?"

"Jeff, I am exhausted and full of all kinds of emotions right now and I want you to know I plan to move swiftly and efficiently on this matter. There will be no reconciliation with Dawn. I have been taken for a fool and will not stand for it. I also plan to stay here for a while with my kids and relatives. This is the place where I can think clearly and be surrounded by loving people. I tried to sleep but couldn't so I called you to let you hear it from my lips."

"Matt, do you want me to put you on the speaker phone so that you can direct us as to your thoughts?"

"No, Jeff. At this point I want your people to investigate Dawn, her brother, and anyone else you or your team thinks should be checked out. In the meantime, please put Damiano on the line."

"Yes, of course, Matt. I will be reviewing the family trust and marital documents throughout the evening."

Damiano comes on the line and Matt instructs him

to keep him posted a minimum of once each day on the progress of Jeff's team.

"Damiano, I want you to call me at the following number when you leave Jeff's office. I have something to share with you that I would prefer not to speak about while you are there."

"Yes, Matt, of course. I will call you back shortly."

Damiano returns to the conference room, where he tells the team that he will need them to send him a progress report by 4 p.m. each day so that he may give Matt a daily briefing. Jeff offers to set up a daily conference call, but Damiano says that will not be necessary. Matt would prefer to hear the information from one person rather than a group on a conference line. Jeff shows some concern over this, and once again Damiano stands and this time does not sit back down. He walks to the door and pauses for a moment, then turns to the group says, "I know you will all do your very best for the Messina family. Thank you for your services and I'll speak to you tomorrow at 4 p.m. I will call your office line, Jeff. Good evening, everyone."

The stage is now set for Matt to be able to plan his strategy through Damiano without Jeff knowing all the details, as this will require some clandestine activities that the legal team should not be exposed to. Damiano returns to the parking garage and heads back out through the midtown tunnel of Manhattan. Once on the Long Island Expressway he calls Matt from the car. Matt picks up after a few rings and asks Damiano if he is alone.

"Yes, Matt, I am still in the car and as always, the traffic is wall to wall here on the expressway."

"Damiano, I want you to listen very carefully and I will explain more as time goes on. I have given a great deal of thought to how I want to deal with my situation with Dawn. It has only been a few days so I need to let my mind settle, but I have a good idea what I want to do. We must get all the information possible from Jeff's team on the legal aspect of things as well as who is involved along with any possible motives. While this appears to be a case of infidelity, I would think there is a component of conspiracy here. It is for this reason that I want my bother kept away from all of this. You are to keep him posted but distant from the management of this crises. He also needs to focus on a business project he is working on for us in Dubai. When you think of what Shamus did to cover for Dawn I don't think he would do that just to protect her from getting caught in bed. He is a simple guy, but there is an odor to this whole thing. I have a hunch this is more than a horny wife."

"Matt, listen, none of us wanted to tell you how we felt about her. You were so happy from the moment you met her, and frankly, I think she really loved you. Your kids and friends were always suspect of anyone in their lives other than dear Rose, but that is a typical emotion as everyone wanted to protect you, and frankly, their own emotions. What I am getting to is that while I am sure she loved you she may love your assets more. So that said, I will do whatever you want me to do to assure whatever plan you are thinking of is well executed. By the way Matt, was Rose surprised?"

"Yes, she was, and that is a story in itself. We had quite the arrival and welcoming crew over here. Listen, I am sure Jeff's team is exceptional and I trust him but make sure he coordinates and collaborates with our corporate director of security with you on point. That brings me to another topic. I am not sure we will be keeping our security folks around, so this is where the plan starts. Contact Jeff in the morning and tell him of my concern with our existing security team. While he must work in concert with them, he must investigate them as well. Surely they must know something of Dawn's activities, so this is a case of keeping your enemies as close as possible. They are very sharp people, but many of them work for us and we need to find out if there are a few bad apples or if the tree needs to be uprooted."

"I see, Matt, I will make that my first order of business tomorrow morning. At the rate this traffic is moving I may be making this call from the car tomorrow."

"Damiano, when this thing is all figured out I want you to come here on a vacation, or anywhere you like. I think you are going to need one."

"Matt, you know I am not one for vacations, but I will take you up on that when the time is right."

"Drive safe and I will talk to you tomorrow at 4 p.m. your time. Please send the kids my best and give Rose a kiss for me. I really miss her."

"Ok, be well my friend. Will do." Matt hangs up the phone and walks over to a window where he sits in a chair as if to look out. He is alone and thinking about the events of the past few days. He reaches down to his foot where a bandage is still on it from having stepped in the glass of the broken picture frame. He rubs his hand over the bandage, back and forth a few times, then grabs an edge and rips over the bandage with a violent tug. He grimaces and says, "This pain is nothing, Dawn." He repeats to himself over and over: "I am not a bad man. I don't do bad things and I will not do bad things. Lord God, please hear my prayers today. I have never experienced the feeling of hatred or malice to anyone I have loved and I don't want to do anything that would take me from your grace. Please help me, God, I need your guidance. I pray that you help me think though my situation and protect not only myself but the ones I love. Lord, I feel that you have brought me to this place so that I may clear my mind and connect with you and family."

As he says these words he begins to realize that God has already put him on a path that may lead to the best possible answer to their situation.

"God, I am really a simple man that has been given so many blessings, and for the first time in my life I feel that the world has betrayed me. I have never doubted you, Lord. I never once complained to you or anyone on earth about my blindness. Yes, I have cried to myself, but you know that I never felt you did this to me. I have always believed that you bestow things upon us that we can handle and that it is our will to decide how these things are dealt with. My blindness has enabled me in many ways, and even at this age I think I am learning more that this is not a disability as many see it. Lord, I not want anything bad to happen to Dawn, but I am very upset and must protect my family. Lord, I make you this promise: I will keep an open mind and pray to you to bestow upon me the capacity to make decisions about this situation that do not cause me to sin but in fact may require me to navigate the trappings of this earthly place. The temptation to do bad has never been more present in my life."

He rolls up the bandage into a ball and throws it across the room. Matt stands and places his hands on the bottom of the widow and opens it a bit. He takes a deep breath of the fresh air coming in the room. There is a gentle breeze billowing the curtains as Matt walks back to the bed where he falls asleep with a confident smile and dreams of his beloved departed Rose and their adventures with the Alucia treasure.

Chapter 11:
REDIRECTION

Matt is in his insurance office on the phone with Rose. She is briefing Matt on what is happening with the government officials.

"So let me get this right. The US government, the State of Florida, and Portugal are all saying that the treasure belongs to them? Rose, what did we get ourselves into down there? How could all those entities actually claim it belongs to them?"

"Well, Matt, it seems that when the Alucia went down it did not do so all at once. The theory is that when it passed over a coral reef it ripped open in shallow water within Florida and then it ultimately went down in open waters. Thus, it makes it confusing, and frankly, Portugal doesn't care where it sank; they say it belongs to them. It is a mess, but Shelton said he had expected something like this might happen and has been preparing for it."

"Rose, this is going to be a long, drawn-out process and we are not prepared for something like this."

"I know; can you come down here, please?"

"I have to tell my brother now. He needs to know what's up. I will speak to him at lunch. Please call me back around 2 p.m. and we will figure everything out."

Matt walks down the hall and makes plans to have lunch with Izzy.

"Brother, I need to speak to you in private. Why don't we pick up something and head over to my house?"

Izzy pulls the car up to the front of the office and Matt gets in and is very quiet. Izzy asks, "Are you ok, Matt?"

"Yes, everything is great."

"So, what do you want talk about that we needed to be alone?"

" It's a bit of a story. Why don't you run into Mancini's Market and grab us some nice deli sandwiches and we will talk at home."

"Sounds good."

Matt waits in the car for Izzy while he gets the usual: prosciutto with capicola and provolone on a fresh Italian roll. The two arrive at Matt's house and head into the kitchen. Matt grabs two bottles of coke and they sit down at the table.

"Well, brother, I have quite a story to tell you. Rose and I met a group while we were on our honeymoon and they persuaded us to . . . um . . . give them some money."

"What do you mean, Matt?"

"Before I tell you that I will cut to the chase and tell you this is likely good news and that they found what they were looking for."

"What do you mean, 'looking for'?"

"Well, Rose and I invested our own money in a search for treasure."

"You did what!?"

"Yes, Rose is actually there right now with the team."

"Team, what team?"

"They have a group of professionals, everything from a lawyer to a historian, organizing things. Rose is on the team as the accountant."

"Huh, wait a minute—she is new to that profession and they picked her? Are you not concerned that these folks are a bit off their rocker?"

"Izzy, I know how crazy this sounds, but this is very real. The governments of Portugal, the United States, and the State of Florida are all laying claim to the find and Rose has to be there to record every single piece of treasure that is brought up from the sea."

Izzy puts his sandwich down and stands up.

"This is a joke, right? Is this is some sort of prank?"

"No, brother. I assure you this is very real. I have to get down there and be there to support Rose. I will go with you, Matt."

"Izzy, we must not tell anyone about this. We don't need the word getting out and having people think we are suddenly rich."

"'We,' Matt?"

"Izzy, you know me. If this is real this is going to be amazing for us as business partners—and more importantly, as brothers."

"All right then, I will tell the folks in the office that you and I will need to go on a business trip that one of our insurance companies is hosting until we figure all of this out."

"What, are you going to tell your wife, Izzy?"

"I will tell her the truth."

"That's good; please make sure she has an understanding of why this needs to be kept very private."

"You know her, Matt; that will not be a problem."

"Yes, she is a very solid person. We are lucky guys. Rose is calling me back at 2 p.m. to let me know the latest and confirm that I am coming. I am sure she will be glad you will be there as well."

In the car ride back to the office Izzy says, "If this doesn't just beat all. I cannot believe what is happening."

"Yes, that's for sure, Izzy. We really need to keep this on the down-low for as long as we can. People tend to go a bit nuts when they think you have a lot of money. Also, people will think we were off our rockers putting money into something like this. That is why Rose and I have kept it quiet. If we lost all the money, we still would have our home and some savings. We figured we were young and could take a risk."

Izzy looks over at his brother and says, "Wow, Rose must certainly have a way with you. You would never have done something so risky in the past."

"Izzy, from the first time Rose and I met we were bonded at the hip and trusted each other's intuition. It feels like we have one thought process between the two of us. We don't feel like two separate people."

Izzy smiles and continues the drive to the office where Rose calls promptly at 2 p.m. Izzy and Matt are huddled by the phone in Matt's office with the door closed. Rose explains that the group is meeting in Dry Tortuga, an island about seventy miles west of Key West.

"Rose, why so far?"

She goes on to say that they are all leaving from separate places in the Keys so that it appears they are not together and this will be far away from the crowds and officials that are so interested in knowing our every move and personal detail.

"Well, that sounds a bit extreme, Rose. All of us are entitled to council and privacy."

"Yes, Matt, that is true, but the culture here in the Keys is a bit crazy and everyone is paranoid so we are all meeting there tomorrow to plan our strategy. None of the crew is invited as they have been the ones who leaked things to the public initially. I need you here, Matt; when can you be here?"

"I am coming with Izzy; I gave him all the details at lunch. We will fly into Miami and drive to Key West. Can you please do us a favor and set up a hotel room for Izzy in the same place as you?"

"Matt, what will you be telling the others up there? The two of you traveling like that together will most assuredly raise concerns."

"Ah, not to worry, Rose, we are telling people that Izzy and I are going to check out an insurance symposium and leave it at that. Stella and the folks will be ok here in the office for a few days. Izzy is going to stay for about three days so we are thinking we will arrive Thursday and he can stay for the weekend and leave me there with you until the dust settles and there is some organization around all of this. I love you, Rose, please be careful down there and send my best to Paul and the rest of the group."

"As Matt hangs up the phone Izzy is shaking his head.

"Matt, I am completely shocked by all of this."

"Yes, brother, that makes two of us; it doesn't seem real, does it?"

"No, I have to say it doesn't, but clearly we need to make sure all your interests are protected and suggest you have someone other than a member of the group down there to keep an eye on things. With all due respect, Matt, from what I am hearing they are flying by the seat of their pants."

"Izzy, I did meet with Jeff a while back to review the investment papers and he gave them his blessing from a legal perspective but he thought we were off our rockers.

"I suppose we should meet with Jeff. Can you do that this afternoon if he is free?

Matt is able to secure an appointment right away and the two drive over to Jeff's law office on the other side of town. Jeff's receptionist greets the two and brings them into the conference room where she offers coffee. Matt asks that a whole pot be nearby as this may take a while.

"Certainly, gentlemen, will do."

With that, Jeff comes into the room and greets the two.

"So, guys, what is the reason for the urgent meeting?"

"Well, Jeff, you may want to get comfortable. Do you remember that investment Rose and I made?"

Jeff slammed his hand on the table.

"I knew it! Those people ripped you off?"

"Well . . . not exactly, Jeff. I thought by now you would have called me because this was hitting some of the new services and I thought you would have made the connection."

"What 'connection,' Matt?"

"Jeff, they found the treasure and were descended upon by all kinds of government officials from the State of Florida and the governments of the United States and Portugal, all claiming a piece of the action."

"Oh my gosh, guys; this is unreal."

"Yes, tell us about it; we are all still in shock."

"Well, first things first. I need the name and number of the lawyer down there. I need to speak to him as soon as possible."

"Izzy and I are heading down there in a few days."

"You should as soon as possible, Matt, but I need to speak to that attorney right away. I don't think they can freeze your assets, but they may be able to make life hard for your investment company down there. I will have my folks pull your file on this and I will refresh myself on the content right away. I am in the middle of a couple of big trust cases right now, so I will bring the file home with me tonight. I suppose this goes without saying, but you are keeping your involvement in this quiet, correct?"

Matt and Izzy reply almost in sync with, "Yes, of course."

"Ok, then, and just thinking back a bit I should have the contact info for that other lawyer in the file."

Matt explains to Jeff that he may have a tough time reaching him as they are heading to a remote location tomorrow to meet and plan down there.

"Ok, I will try to reach him before I leave today. Matt, I have known you a long time now and I must tell you this still amazes me that you got involved in this thing. Rose must certainly bring an interesting dynamic into your life."

"Jeff, its hard to explain but I trust her every instinct."

"Well, Matt, she did pick you and that confirms her good judgment in my eyes.

"Jeff, yes indeed, and in my 'eyes' as well."

The three men have a bit of a laugh at the remark, shake hands, and head out of the conference room. As they pass Jeff's secretary on the way to reception, Jeff requests the Alucia papers from the Messina file. Jeff walks back into his office and with a huge smile on his face says to himself, "Well doesn't that just beat all . . . that just beats all . . ."

Jeff calls Matt the next morning in his office and advises him that there are no apparent personal exposures other than the possible loss of their funds if the government officials are successful in their seizure. He also tells Matt that he was unable to reach the other attorney and asks that once they get to Key West to please make sure that the Alucia group's attorney reaches back out to him. Matt thanks Jeff and tells him that they will be there in two days and will be certain to keep him posted. As the day progresses Matt and Izzy are busy taking care of things in the office to prepare for their trip when Rose calls. Matt is quick to tell her that Jeff advised her that it is not likely that they have any personal exposure beyond the possible loss of funds and that Alucia's attorney needs to call Jeff to discuss this in more detail.

"Sure, Matt I will have him do that."

She goes on to tell Matt that the group has determined that what has been found is still likely to be the mother lode but the ship's manifest would seem to indicate that there is far more to be found. They were also able to confirm that it truly is the Alucia from markings on silver bars as well as the cannons.

"Matt, this place is crawling with reporters and treasure-hunting nuts."

"Rose, what is being done to prevent looters and such out there?"

"Well, honey, the Feds and State Police are on the docks as well as patrol boats and everything is under protection while every item is being archived and recorded. I am playing a significant role in that process. They have counted five hundred silver bars that weigh eighty or so pounds each and there are far more down there with literally thousands of gold and silver coins."

"That is crazy, but I still don't really understand why you all had to go out to Dry Tortuga to have that meeting."

"Well, we all felt it was the only place we could speak in this region without the walls having ears."

"Wow, it must be crazy down there."

"Matt, you have no idea."

"Please tell me you are being careful."

"Matt, we are under police guard at all times. I am very safe here."

"Ok, Izzy and I will be flying down to Miami tomorrow and we will grab a rental car and drive the rest of the way."

"I cannot wait to be with you, Matt; this is beyond overwhelming."

"I am looking forward to being there as well, Rose."

"Matt, they are saying this is going to be a long legal battle and we have not even gotten started. We cannot sell the treasure to keep things moving. Legal fees and day-to-day expenses are mounting, so we are considering selling more shares of the expedition. Offers are flying in, so there is much to talk about."

"Well, I guess that means that the prospective investors really believe that the expedition will win this fight, Rose. I would think you had better get another leagl opinion of selling shares in the midst of all this. Perhaps it is just a matter of complete and utter full disclosure."

"Yes, we are really going to need the money for legal fees and to keep the dive team paid, not to mention keeping our equipment running every day. This is a real interesting situation. It's like winning a huge lottery without the ability to cash in."

"Rose, we will prevail, I am sure of it. I guess the game will be how much of the expedition the group sells to investors."

"Yes, Matt, that was one of the topics we had discussed at our meeting."

"Well, seems like there will be lots of complex issues to deal with. I have to run, Rose. Izzy and I are working hard to get the office work done so we can leave it in good order. We have kept this from everyone and they are all convinced Izzy and I are going on a business trip. No one has connected us with this and Jeff has made it clear we need to do so for a while. I love you, Rose, please be careful and see you soon."

Chapter 12:

STICK BALL

Matt awakens in Palermo from his sleep to the soft breeze coming through the open window. The air is filled with the smell of fresh baked bread coming from the kitchen downstairs. He swings his legs off the bed and pauses to feel the sunlight on his face. Matt takes a deep breath and walks over to the window where he sits in a chair facing out to the street across the square, which is where they had the welcoming party the prior day. The sound of young children playing stickball can be heard and Matt recognizes the voice of one of them as young Salvatore. Matt stands and places his hands on the windowsill and listens intently to the game, which reminds him of his days back in New Jersey with his brother and his friends. A crisp crack of the stick is heard and the ball comes close to the house where Matt is listening to the game. Salvatore looks up at the window and sees Matt standing there. Matt is unaware of the fact that Sal is standing below him at this moment, but can tell someone has come to get the ball. It had made its presence known with a familiar thump on the ground outside the window when the ball landed and he had heard the fielder coming after it. Salvatore stops below the window and stares up at Matt while his friends are yelling at him to pick up the ball and make the play. The other boys are furious. Salvatore picks up the ball and throws it to the closest player and waves to the others that he is not playing any more.

Matt knows he is there now and asks him if he made the play. Salvatore is amazed by the remark and he asks, " What do you know of this game, sir? How did you know what was happening?"

"Son, I love that game and used to play it as a young boy as well. I heard the ball land over here after it was hit."

"Sir, please? How can a blind man play this game?"

"Salvatore, I have a ball that helps me. My mother gave it to me a long time ago. I keep it as a good luck charm and have it with me in my suitcase."

"Will you show it to me, sir?"

"Yes, under one condition: you allow me to play with you boys."

"It would be an honor, sir."

"I will be out in a few minutes."

Rose knocks on the door and asks, "What is all this commotion, father? Who are you talking to out there?"

"I am going to play stickball with the boys outside."

"Father, please come eat; we have made an amazing breakfast."

"Honey, I will eat soon, but I have an important date to make. I need to get changed."

"Ok, father, but please, we have worked hard so don't be too long and I will get Luke and Mark to join you."

"Please, Rose, don't bother them if they are sleeping."

"Yes, they are."

"Ok, then, let them sleep a bit; they have had a long journey these past couple of days and we will all eat soon."

"Have it your way, father, but I am coming, too."

She closes the door and Matt gets his special ball and changes quickly.

He has a T-shirt and suspenders on with dress pants and sneakers. This is quite the sight and not the typical outfit for such an event. Matt's hosts see him come down the stairs with Rose and they say, "Matteo, you never cease to amaze us." They laugh a bit and all walk out where Salvatore is waiting to greet him.

"Good morning, sir, can you show me this special ball?"

Matt reaches into his pocket and a jingling noise is heard. Matt pulls out the ball and it has a bell inside it that is sort of a ringing rattle.

"You see, Salvatore, my mother had this idea when she was in a park one day. She observed a woman playing with a dog, playing fetch with this same type of ball. This was made to entice the dog to fetch it and be able to track it once it was thrown. So, my mother thought that I could hear the bell and hit it with the stick when I was a batter as well as being able to follow it on the field after it was hit. Every now and then I would even catch one."

"Sir, I must see this, but please forgive me for a moment; I want to explain this to my friends."

"Sure, Salvatore, you go ahead; I want to be sure they welcome me to play."

Rose, standing by, says, "This is too much, Father. Are you sure you want to do this?"

"Honey, you have no idea how much I want to do this."

"Dad, I remember Uncle Izzy telling us about some of these games. I bet you guys had lots of fun."

"Honey, there were some days when we played from sunrise to sunset, and after the sun had set we still wanted to keep going but at that point I was the only effective player."

They both laugh at that remark. Salvatore is over on the other side of the square and the boys keep looking over at Rose and Matt as their discussion progresses. They don't all seem to be embracing the idea. Matt tells Rose to walk him over to the group. Matt says to Salvatore, "There seems to be some debate here. Salvatore, take my ball and give me the stick. Throw me a few pitches."

The others stand back. Salvatore takes Matt by the hand and walks him over to where home base is marked on the cobblestones. Matt taps the bat against the ground and takes a few practice swings. No one steps up to be the catcher, as perhaps they were concerned about being hit by the bat. Salvatore takes his position on the mound and asks Matt if he is ready.

"Yes, ready when you are."

With that, Salvatore throws the ball with a gentle underhand motion and the ball lands short in front of the plate and bounces behind Matt. Matt tracks the ball and throws it back to Salvatore while saying, "You don't have to worry; throw the ball like you normally would." The boys standing to the side watching this are becoming curious and one shouts, "Wait! Let me be catcher." The others stay on the sideline and watch. Salvatore throws a second pitch and the ringing ball tracks straight down the middle, right over home base, where Matt takes a swing and misses it. The catcher is there and throws the ball back. The boys on the sideline are quiet and wait for the next pitch. Salvatore asks if Matt is ready, to which he replies, "You don't have to worry or ask; I will hear the bell as you wind up and then when it is coming."

Salvatore winds up and unleashes the pitch and with a crack of the stick the ball is sent flying farther than the young boys have ever seen. All the boys cheer and Rose yells, "Great hit, father, great hit!"

By this time, all the commotion has woken up Mark and Luke and they come out to see the game. The three siblings stand by to watch Matt hit a few more times and with each hit more of the doubting young players come into the street to take positions. Rose and her brothers are in awe of the moment. Rose says, "You know, I don't think I have seen a smile like that on Dad's face since mom was alive. I have forgotten how his face seems to change when he smiles like that. He looks like a different person."

"Rose, We have seen a similar smile over the past couple of days, but nothing like that."

"Hey, we have to play; what do you think?" asks Luke.

"Oh yeah, for sure, lets go in and get our stuff."

By this time their hosts realize that breakfast is going to get cold so they bring it out to the square and in true Sicilian fashion there is enough for everyone to eat, even the boys playing the game. No one can resist the fresh bread and cheese. Everyone takes a break to enjoy the meal. As they are eating, boys from the village ask Matt different questions about his condition. One says, "Sir, you must be able to see a little, yes?"

"No, young man, I cannot see anything."

Another asks, "Do you see darkness?"

"I see nothing; it as if I see with my ears, mind, heart, and other senses like touch. It is hard to explain, but I was able to see until I was nine years old, so I understand things like colors, houses, trees, mountains, people, and all the things we may take for granted."

As Matt continues to speak about his blindness with the young men, Salvatore's mother walks up to listen to the discussion. She has been watching the stickball game from a distance. She, too, has brought more to eat and drink for those playing and gathered in the square. Quite a crowd has come out by this point and dozens of people are now partaking in the fun. Rose taps her father on his shoulders and says she wants to play with the boys. This is not something that goes over very well with the young men, but Matt intervenes and says, "Surely you boys are not too worried about being schooled by a girl, are you?" What the others don't know is that Rose was a high school softball champion and played in college as well. The boys are very reluctant. Matt says, "Ok, then, how about a blind man, his daughter, and his sons play on one team against the others. Salvatore is quick to add, "Sir, I want to play on your team as well!"

The other boys laugh and say, "Of course. This is going to be an easy win."

Salvatore's mother gives him a nod and wink of support for his selection. A bystander offers to flip a coin to decide who will bat first. As the coin is in the air Salvatore yells, "Tails!" The coin lands heads up. The Messinas take the field with Matt at third base, Rose pitching, Mark at first, Luke at second, and Salvatore to cover the entire outfield. Rose calls out to the players to have a quick meeting at Matt's third base.

"Ok, you guys remember how we used to do this on summer vacation? Dad, you try to field the ball if it is hit to you and no matter what I will call out where you are to throw it. So, if it is to Luke, I call it, then Luke you shout to give dad your location. If dad has to tag the runner coming to him we roll the ball on the ground and dad will trap it and make the play that way. Salvatore, do you understand?"

"Yes, I do; I am ready."

The excitement in the crowd is palpable as its size continues to grow for the game—and of course, the food. Rose rolls a few balls to Matt and he throws a few back to her to get oriented on the field of play. Luke and Mark do the same. Salvatore being out in the field stays out of the warm up and watches in awe. The boys getting ready to bat are beginning to think that they may be up for a challenge but are confident that their win should be easy. A local boy not playing volunteers to be the steady catcher for both teams as they are short one player each and another is selected to run for Matt as base runner should he make a hit. Rose's first pitch is met with a swing and miss. The batter turns to his friends and says not to worry; she has was lucky with that one. The second pitch is hit far into the outfield, where Salvatore has to chase the ball. Rose yells, "Throw it to me!" As the runner begins to pass second base, all realize that the play is going to have to be made by Matt. Rose catches the ball and yells to Matt, "Coming to you, dad!"

She gives it a rapid roll and the ball is rolling and skipping across the cobblestones to Matt, where he puts both hands down to touch the ground and the ball hits his left arm as he fumbles for the ball but does manage to get it. The runner realizes he is going to be tagged out at third base and backs up to second. There is a huge roar from the onlookers and the boys at bat cannot believe what they have just seen. Salvatore's mother has her hands raised to cover her mouth and is overwhelmed with the sight. The game goes on for a long time with the lead changing back and forth for what seems like hours. They play through the lunch hour and are now in the last inning. The game is tied and Matt is at bat with two outs, so the game is up to him.

By this time there are so many people in the square there is nowhere left to stand. Some are in windows, others are standing on ledges and along the fountain in the middle of the square. The cheers could be heard blocks away. Even the local police have come to watch the game. In a show of sportsmanship, the pitcher calls for silence, as he knows Matt will have trouble hearing the bell in the ball with all the noise. The crowd complies with his request and he releases the second pitch. Rose is on third base as the only runner and she is crossing her fingers behind her back and watches as Matt takes a swing and a miss.

This is it, the last pitch, and the only chance to win the game. The crowd is cheering even louder and this time it takes the pitcher a bit longer to calm the noise. The pitcher looks around to his players and watches of Rose's location one more time. The pitch is sent to Matt and he cracks a great hit right over the pitcher and second baseman. The ball takes a wild bounce into the fountain and Rose tags home plate. It may as well have been the World Series. Matt knows Rose has crossed home plate and they have won the game. The cheering went on for quite some time. While all the hugs and well wishes were exchanged, the young boys who had lost were very gracious and said they would never forget this day.

Salvator's mother breaks through the crowd and takes Matt's hand and places the now wet ball, which she had retrieved from the fountain, and in his hand. She is without words. Matt feels her silky, smooth, warm skin and smiles and says, "Thank you, young lady." She blushes and stares at him. She is overcome by his touch and overwhelming good looks.

"Thank you, sir."

Salvatore can see a look in his mother's eyes that he has never seen before. Everyone standing close by is affected by the tender moment. She thanks Matt for the compliment and for letting her boy play on his team and comments that the village may be changed forever. She comments that many lessons were learned by the villagers, especially the young boys whose minds were opened by playing a game reserved for men and for understanding more about people with disabilities.

"Yes, my Rose is a very strong woman in many ways, just like her mother was—God rest her soul. I should be the one thanking you and the others for the great time I had; I have learned things as well. It made me recall my youth and the great times I had back then. I am a man who has many things and great privileges and nothing comes close to being able to be with family and friends. There are no greater riches than the love of family and friends."

Salvatore's mother says, "Yes, I agree, and after my

husband's passing everyone in this village took care of us and to this day they have never stopped helping in many ways. I work at the church as the secretary to the parish priest. You must meet him. He is a very wise man."

"I will be going to church on Sunday and would like to meet him."

"Even better: I cook a meal for him every Saturday evening. Would you and your family like to join us?"

Rose, standing close by, says, "Yes, we would love that," before her father had the chance to reply.

"Dad, I know you will love Father Passante."

"Ok, it is settled, but please let us bring the wine and some desserts."

"That will be fine, sir."

"Please call me Matt, and my boys are Luke and Mark."

All standing around sense the closeness and warmth of a friendship that is about to blossom.

"What time would you like us to be there tomorrow?"

"Father Passante is like a clock; he will be there promptly at 5 p.m., right after he does his afternoon rounds to bless people in the hospital. Salvatore will come to get you about fifteen minutes before. Our flat is a bit hard to find, so it is best if he comes to walk you there."

"You are very lucky to have such a nice mother, Salvatore, and we are all looking forward to being there."

The families exchange some hugs and go on their way. As they walk back to the house Rose thinks to herself about the exchange and sees a part of her father that she has not since her mother passed and becomes even more curious about Dawn's absence. Once they get home everyone goes back to their rooms to get cleaned up and changed. It has gotten to be late afternoon. After Matt settles in he goes back to sit by the window and again relaxes for a few minutes with the ball in his hand.

Chapter 13:
WINDOW TO THE VILLAGE

He begins to clench the ball with a very tight grip, over and over again, as he thinks about what Dawn has done. He walks over to the phone to call Damiano, who picks up the call in the kitchen of the Hampton home.

"Damiano, my friend, I am calling a bit early today because I want you to do something for me."

"Sure, are things going well over there?"

"Yes, they are better than I had even hoped and I will tell you all about it another time. I want you to call Dawn and tell her that I will be out of town for an undetermined period of time and confirm that she is not to come to any of our properties other than the place in Jersey City. Let her know all of the accounts have been locked other than the joint operating account, which will have a few thousand dollars put into it each week. She is not to contact any of my children and is welcome to come to the Hampton home to pick up her things only if you escort her through the house and watch her every move. I don't want any of her pictures or those of her with any of us in the home when she comes, and she must come alone. I will likely not call the rest of the weekend; call me back to let me know if she does not agree to any of these things. We have a nice weekend planned and I really don't want to have to deal with this at the moment if I don't have to. I have been thinking a lot and have some plans that I will share with you soon as it relates to Dawn."

"I will take care of this right away, Matt; give my best to everyone."

Damiano calls Dawn right away and makes sure she understands. She knows better than to give Damiano a hard time and agrees to everything he tells her. Before she hangs up she tells Damiano to please forgive her and says that there is more to this than he knows and that she will come to the home Monday afternoon.

"I have nothing to say to you, Dawn—certainly nothing that a gentleman would say," and he hangs up the phone.

Back at the village, the Messinas are gathering for dinner with their hosts, the Bianca family. Giovanni is helping his mother, Constance, set the table as the

Messinas make their way into the living room next to the dining room. Mark, Luke, and Rose all offer to help and are each given a task to do to help get the dinner ready while Mr. Bianca and Matt walk out to the courtyard behind the house to get some alone time. Mr. Bianca is a very proud man who doesn't say very much but has great instincts. He runs a business that has been in his family for many generations and their family has never left the village. Their ancestors lived through many challenging times, like World Wars I and II, but always managed to keep the family and business together.

"Matt you must not let such a long time go between your visits here."

"I know, Joseph, we have already spoken about that as a family and I am sure we will come more regularly. We would like you to come to visit with us in the States. We have more that enough room for everyone and there is so much to see there."

"I can just imagine what your homes are like Matt; we are very happy for you and your family, and that is very nice, but it is so hard for me to leave the business and something like that is too expensive for us."

"Joseph, please don't take this the wrong way, but I want to make it our treat."

"Matt, that is very kind of you, and I don't take it the wrong way. I know you have the best intentions and if I had the resources I would do the same for you."

"I know, Joseph, please speak to Constance and Giovanni when you think the time is right. The offer stands, and in fact, it may be a better idea than you think. It is hard for me to get away from my business, too, but from now on I will be making the time, no matter what."

"Matt, perhaps you are right, and this may be a good time to have Giovanni stay and run the business without me looking over his shoulder. I know he has bigger aspirations than to run a family food market, but you know yourself, Matt, you cannot run a big operation well until you have run a small one."

"Joseph, that's the spirit. I hope you and Constance come soon. I am sorry to hear that Dawn will not be here for this visit."

"Joseph, I have something to tell you, but you must not share this with anyone other than Constance. I am going to divorce Dawn. She has been with another man in our bed. I am not sure how long it has been going and I have already started the process of getting her out of all of our lives. The boys know but I don't want Rose to learn of this. I am concerned that she would leave to come home if she knew about this and I want to be sure she finishes her studies. She does not worry about me right now. I have many people looking after this situation and I will be making a daily call to the states to get updates."

Joseph stands up from his chair and places his hand on Matt's shoulder and asks Matt if there is anything he can do for him.

"You and Constance have done more than you know by having us here. I do have one request: does Constance still keep bottles of the Baby Amarone that Rose and I used to love so much?"

"Oh yes, we have many. I will go to the wine cellar and fetch some now."

While Matt is waiting, Constance comes out with a board of cheeses, olives, and prosciutto.

"Matt, your children have grown up to be fine young adults and I must say that Rose reminds me of your dear wife, Rose."

"Oh yes, Constance, she is very much like my Rose: strong, intuitive, free-spirited, and very smart. Rose and I taught them to stick together and be humble and to develop servant's hearts."

As Matt is saying this, Joseph comes back with the wine and agrees that those are essential qualities.

"Our Giovanni seems to be finding his way."

"Yes, he is; we had a nice discussion in the car on the way here."

"I will leave you men alone for a while. The kids and I are having a nice time preparing for dinner and catching up. Joseph, do you know the young boy, Salvatore, who was playing the stickball game today? Matt, he and his mother have been through a lot. Her husband was injured in World War II and struggled with his health until he passed. It was hard for him to hold a job, so the mother took odd jobs, and many people here in the village have been helping her and Salvatore since the husband's death. The husband was a bitter and proud man. He would never accept help from anyone, not even a gift of a loaf of bread. He made their lives even more difficult, but they never complained about him and served him faithfully as the head of the household until he died. Salvatore and his mother visit the grave every Sunday after church and bring flowers there."

"Joseph, I took it upon myself to invite them to our Sunday celebration this weekend as they invited me and the kids to their home to have dinner with the local priest tomorrow."

"That is great, Matt, and you will enjoy Father

Passante; he is a rather special man. He has done much for the village and its people. In fact, he will be here for our Sunday celebration," Constance says. "Joseph, how long has he been the parish priest?"

"He has been here since late in the 1960s; I think it was 1969. He will likely be elevated to Monsignor very soon. He is on the board of the bank here in the village and has been key to rebuilding our school and town infrastructure. The decade following the war, this place saw a very slow recovery."

"I recall some of this, Joseph, but never knew all this details about the village. The kids were so young last time we were here and it seemed we were busy with the kids and did not learn very much about things here."

"Do you have plans to take the kids to your parents' old home?"

"I would like to do that, yes. Do you know who lives there now?"

"It is the same occupant, Mr. De Francesco . He is in his late eighties now, however, and his wife died some years ago. He still wears a fine suit every day and walks to early church service and never misses Sunday Mass. He is quite the fashion statement, from the tip of his walking cane to brim of his Fedora. His is a very proud veteran and flew the Savoia-Marchetti plane in the Spanish Civil War and then for the Luftwaffe as a volunteer. He tells the same stories over and over again, but we all listen and honor the man. I am sure he will let you come to see the home. One trick, though: be sure you bring him a fine bottle of Scotch. He will love you for it."

Constance comes out to let them know dinner is served. The men join the others at the table. Matt is asked to say a prayer and is honored to do so. All join hands, close their eyes, and bow their heads as Matt says: "Father, we are thankful for this food, this day, and these people, and pray for those that serve our lands and protect freedom-loving people around the world. We may have had political differences but we all still honor thee as our God and pray for peace and love throughout the world. Please protect all that love thee and forgive those that have drifted from your ways, as they know not what they do. Help us all to forgive those who sin, Amen."

As Matt is praying, Joseph raises his head, opens his eyes, and looks at Matt, thinking about the inner torture Matt is experiencing because of Dawn and the fact that he is certainly praying to for God for strength to forgive her. After the prayer is over Rose says, "That was a good message, father."

Constance agrees and adds, "Forgiveness is a virtue; it helps us rid our souls of unhealthy thoughts so we may live better lives in God's way."

Giovanni is quick to say, "This is great; now I will pour the wine, as the Lord Jesus certainly celebrated in this fashion."

Giovanni's remark lightens things up a bit.

Luke says, "Dad, do you have any adventures planned for the next several days? Within the last several days we saw you big game fish, drive a boat, fly a plane, drive a car, play stick ball—"

Rose says, "Dad, what the heck have you been up to?"

"Rose, my love, I have put off a lot things in my life and I suppose I realized it's time to live a little and enjoy a few things that I always wanted to do. Admittedly that is the second time I 'drove' a car. Planes and things that fly have fascinated me since I was a child, and I heard the Hindenburg fly over my school's rooftop as a young man. Our family has had so many blessings, and to be able to own things like boats and planes is a real privilege. Mother always used to say that we must be wise stewards of the things that God brings into our lives and I feel that over the last several years we have begun to forget those words. We need to start giving back more to society. There are times that I feel guilty about all that we have because there is so much suffering on this earth. Yes, we donate to charity from the business, but I would like to see us be more involved with causes on a personal level."

"Dad, how about we go around the table and each share which philanthropy we would personally like to get involved with?" Luke suggests.

"That is a great idea, son; would anyone object to this?"

No one does, and Giovanni asks to go first. He tells of his love of the sea and that he would work to help keep the waters around their island safe and clean so that the fisherman could be assured that they had an ongoing supply of food to supply to the island, which is a major part of the economy and important to ecosystems. Rose says she would like to be more involved in the fight against cancer to honor her mother. Mark and Luke concur and say that is exactly what they would do. Matt says, "Well, I must say, that is what I had in mind, and I think it's time that we set up and run a foundation to honor your mother."

Constance says, "That is a wonderful and beautiful idea, Matt."

It is Joseph's turn, and he says he had an idea but

feels that he would like to help with that as well, as there have been many other friends and family members affected by cancer as well.

Giovanni says, "Yes, I too love this idea," to which Matt replies, "Don't worry, Giovanni, this was not a vote. If you are passionate about the sea you should pursue it."

"Yes, uncle, but cancer is affecting more people around the world and I loved Aunt Rose so much and want to be a part of this."

Matt reaches for his wine glass and stands.

"I would like to propose a toast."

All take their glasses and rise.

"In the memory of our dear Rose, I hereby announce the formation of the Rose Messina Cancer Foundation."

Glasses are clinked and conversation ensues about what the foundation may do to help in the fight for cancer.

"I also propose that we change the name of the foundation that your mother chaired for the disadvantaged people of Haiti and name it after her."

As the evening wears on and the dessert is brought

out, Matt tells the family that he and Joseph were speaking and that it would be nice for them all to visit the house his parents lived in before coming to America.

Luke says, "That would be great, dad; we saw some pictures in the album in the plane and I think we had some at home."

All are quick to agree.

"Great, we can do that tomorrow morning," says Matt, "and we have dinner plans to meet Father Passante over at Chiazano's tomorrow afternoon. I hear she is a wonderful cook, and Father Passante is an amazing man."

Luke says, "We had better get some exercise in, or we will blow up like balloons from all this eating."

Matt says not to worry. "We are going to do a lot of walking on this, trip I assure you."

"Hey, we might want to play more stickball; that will keep us in shape," says Mark.

All laugh and continue with the desserts and enjoy a quiet evening at home after yet another long and exhausting day. The families have grown close in a very short time and the feeling of togetherness and warmth of loved ones brings comfort to everyone. Constance says she is sorry Dawn could not make the trip and that her friend is ill. At this juncture everyone at the table knows except Constance and Rose that this is a lie. Matt says that he will be checking in on Dawn over the weekend to see how things are going. "It must be hard to be away from you wife," adds Constance. Matt chokes a bit on his wine and says, "Well, yes, it is but I have my family so it makes it so much easier to deal with."

Rose looks at her brothers and starts to sense something is a bit off but lets it go.

"I don't know about all of the rest of you, but I am exhausted and will be heading to bed soon," Matt says. "Tomorrow is going to be a long day, so I think I will head upstairs and relax."

Matt kisses the ladies at the table, gives a hug to the men, and heads upstairs while the others start to clean the table and mill around the kitchen.

Rose takes Mark and Luke aside back in the pantry and asks if they think something is wrong with their dad. Mark is quick to add that he thinks that it is emotional for Matt to be in this place and that the last time he was here was with mother. Luke agrees, "Yes, it must be that."

Rose pauses and looks out the window and then back to them and says, "Yes, you must be right. This has to weigh heavy on his heart."

Giovanni joins the discussion and says, "Why all the serious faces here?"

Rose tells him that they were a bit worried about their father but feel it's just a bit tough on him as he remembers their mother, Rose, being here the last time.

"Yes, I see. Hey, why don't we go into town this evening and I will show you a bit of Sicilian nightlife?"

Rose has not been out at night as she has been busy with her studies. She is not the type to hang out in clubs but feels it will be fun for the boys. They all agree, so Rose heads up to tell her father they will be out the rest of the evening. She knocks on the door.

"Father, we are heading out to town with Giovanni this evening and I wanted to say goodnight before we left."

"Sure, Rose, please come in." Matt is seated at the desk near an open window and he says, "You know, nothing smells like this place. Even the air outside smells like a bakery and my mother's kitchen."

She leans over to kiss her father on the head.

"I love you, dad."

"I love you, too, Rose. Please be careful with Giovanni. I think he has become a good young man but I sense there is still a wild streak in him."

"Sure, father, I agree, and this is the first time I am out on the town. I feel much better with Luke and Mark going."

"Very good, Rose, stay close and enjoy. Call a cab if he gets tipsy and wants to drive. Of course, Father, sleep well."

After she leaves, Matt lies on the bed, falling asleep to the sweet smell of the baking bread and thoughts of his mother and brothers in their kitchen when they were young.

Chapter 14:
PANIC

Meanwhile, back in New Jersey, Dawn is in a panic. She is now in the condo and feeling that Matt's team has essentially imprisoned her in a way. She is stomping back and forth, lighting cigarette after cigarette while sucking down martinis. She says to herself, " He cannot do this to me. That son of a bitch cannot just have his army of lawyers and goons lock me out of my life. I am entitled to more than this." She becomes completely drunk and calls her brother, Shamus.

"Hey you, dumb ass. Get over here right now. We need to talk. If it were not for you we would all not be in this situation."

"Well, sis, if it were not for your cheating and scheming this would not have happened. Don't put this all on me."

"Screw you, Shamus, I did a lot for you and will

continue to do so because you are too stupid to help yourself. Now get over here."

"Please don't me mean to me; I cannot take that."

"Ok, Shamus, calm down; I have been drinking and I am pissed off. Please come, we have to talk."

"I will come if 'he' is not going to be there."

"Shamus, I am sure they are watching every move I make, and no, 'he' is not going to be here. Shamus, please pick up some vodka and cigarettes on your way over here, I am low."

"Sure, sis, would you like me to get you some tampons, too? Sounds like you may be having that time of the month."

"Shamus, you ass, just do what I tell you."

"Sure, sure, whatever, I will be there in about an hour."

Dawn calls a lawyer friend that Matt and his associates do not know.

"Hey, Renee, how are you? This is Dawn Messina. It's been a while; hope all is good by you."

"Yes, Dawn, I am well. Nice to hear your voice. What do I owe the pleasure of this call?"

"Well, I have a situation. My husband caught me

cheating and has locked me out of all our properties and even given me limited access to funds."

"Wow, that is a mouthful, Dawn. Did you guys have a prenup agreement?"

"Of course we did; he had so much prior to our marriage and all that goes with their children that Matt did everything he could to assure their legacy was preserved."

"Well, Dawn, did you contact the attorney who drafted it?"

"No way, Renee, I don't want to speak to him. He is one of Matt's active staff lawyers. I didn't have any money to do it on my own and it seemed clear to me."

"What did the infidelity clause state?"

"I am supposed to get five million no matter what the cause of our divorce is. So, basically, if I cheated on him I am not entitled to anything more than five million. I could not go after any of his assets in that case."

"Well, Dawn, it sounds like you are going to land on your feet, so why the concern?"

"Matt must be up to something based on his reaction to this."

"Dawn, come on now. He is a man of great means and resources. Are you telling me the entire story?"

"Well, there is a bit more, Renee. My boyfriend is a senior accountant in one of Matt's companies. He has been siphoning money to a Swiss account for the last two years, using fake vendor billings and the like."

"Oh my god, Dawn; you are crazy. What were you thinking?"

"Renee, I never really loved the guy and I have to admit I married him for the lifestyle. I never thought I would do this to him. I met this guy at an office party and fell for him immediately. We started to meet at different places and over the last two years we got a bit too bold. Not only was he stealing from the company, we were even having relations at our home on Park Ave. My idiot brother was watching the door, but messed up. The buzzer was broken on the day Matt decided to come home early and surprise me with roses."

"Dawn, you are in a very complicated situation. You need to come into my office on Monday. No more conversations on the phone, please, Dawn. Not with anyone until you speak to me on Monday."

"Not to worry, Renee. This place is not bugged."

"Dawn, you do not know that. I will have someone

come over today to go through the place. Will you be there all day?"

"Yes, my brother is bringing me some things and I plan on being here."

"I think this may go without saying, but do not have your boyfriend come there."

"All right, Renee, I will let your guy in. Thanks, and see you on Monday."

She hands off the phone and continues to pace back and forth, spilling her martini as she looks out the window across the Hudson River to the New York City skyline. Soon, Shamus arrives at the condo. The doorman knows him and allows him to pass to the elevator, which opens directly to the vestibule of the condo. He rings and knocks at the door. He is not immediately acknowledged, as Dawn has passed out from all the booze. After some louder bangs at the door she eventually pulls herself together and manages to come to the door to let her ever-faithful brother in.

"Dawn, what are we going to do? I have no job and you and your boyfriend must be in lots of trouble."

"Shamus, shut up; you are not to speak of this."

"But, we are alone, why should that matter?"

"You don't understand, brother. You will slip up and speak to people you should not. So from this moment on, unless I speak of it, you are to say nothing. In fact, your phone and other places may be bugged. I have someone coming to check this place out. After he is done here I will send him back to your place. You tell him nothing, Shamus. His only job is to look for listening devices and bugs of any kind. Have you mentioned this to anyone?"

"No, sister, I just told my landlord that I was home a lot since I lost my job. She was a bit surprised because she knew I worked for family but I told her it was a temporary thing due to some changes."

"Well, that sounds all right, but no more speaking of this for any reason."

"I understand, Dawn, and I am scared. Matt is so powerful and has so much that he can do to us."

"Shamus, we have rights and all will be ok. You must not panic."

"Well, then, you should not drink so much."

"Shut up, Shamus; don't tell me what to do."

Shamus starts to cry and walks out of the room into the kitchen to put down the bag of booze and cigarettes he brought her.

"Hey, brother, bring me the vodka you brought."

He comes back into the living room and hands it to her, where she shakes up a fresh martini at the wet bar.

"Damn, I am out of olives."

She throws the shaker into the sink where it makes a loud clang. Shamus is nervous and uneasy. Afraid to say anything more he just watches her drink and speak to herself. Over and over she repeats, "I am not a rat in a cage. I cannot be treated like this." She passes out in the chair with the martini glass still in her hand. Shortly after this, the doorman announces the arrival of the man sent by Renee. Shamus shakes Dawn's shoulders and lets her know of the arrival and the man is cleared to come up. He is greeted at the door by Dawn and he advises that he is the man sent by Renee.

"My name is Miller."

"Ok, then, Miller, please do what you have to do. I am Dawn and this is my brother, Shamus."

"This is going to take some time. If you don't mind, I must have access to every area in the place—even locked secure areas. Are you ok with that?"

"Yes, of course, Miller. If you have time, can you please follow my brother to his place and do the same?"

"Yes, of course."

It doesn't take long for Miller to find a bug in the phone.

"Dawn, how long have you been here and have you had any discussions on this line?"

"Yes, oh my god, what is that in your hand?"

"It is a listening device."

"I had only two discussions: one with Renee and one with Shamus. Both were today."

"How long have you been at this residence?"

"I came just yesterday and my damn husband has basically sequestered me here since he and I have had some issues."

"Well, he clearly did that for a reason. Please, no more discussions, even in this room, until I complete my sweep of the place. If it is convenient for you to be somewhere else you may want to consider it as this is not a perfect science. I am the best in the business but cannot give you a one hundred percent assurance that I will discover all the surveillance devices."

After a couple of hours of pulling the entire place apart the only device that is located is the one that was in the phone.

"I will head over to Shamus's place now. My car

is around the corner. Would you like to ride with me, Shamus?"

"Sure, Miller, I will. Dawn, I am scared. Will you come with me?"

"No, Shamus, you will be ok. Go with Miller and he will take care of you. I am going to go to a pay phone and make some calls. Miller, please bring Shamus back when you are done. He is my only family and I have to help him out."

She gives Miller a wink as he picks up on the fact that Shamus is a bit slow. The two men leave and Dawn gets her act together fairly quickly. She takes a shower and changes into a designer exercise outfit and heads out to a local coffee shop where she calls her boyfriend.

"Hello, this is Ken."

"Sweetheart, it's me. I am calling from the coffee shop around the corner from the condo."

"Why wouldn't you just call me from the condo phone?"

"Prepare yourself: Matt had the place bugged and I am not sure how long it has been there."

"Huh, what?!"

"Yes, he had a listening device placed in the phone.

My lawyer had this guy come by and check the place over but he said since he found one device that I should take no chances. In fact, he is heading over to Shamus's place right now to see if there are more there."

"Good Lord, Dawn, this is going to be bad."

"My lawyer thinks I need to distance myself from you because of what's been going on with the money but she doesn't know the extent of it. I have a meeting with her on Monday. Ken, we are in this together and I am not going to leave you, no matter how much Matt tries to wear me down."

"Dawn, we need to meet right away and discuss this. For God's sake, we could do a lot of jail time for this. Have you been drinking? You are slurring your words."

"Yes, I am a mess. Before I found the bug I told the lawyer about you and what was happening."

Dawn starts to cry and drops the phone receiver. She picks it back up.

"This couldn't be any worse. Matt and his machine are going take us down fast."

"Dawn, listen to me carefully. We have to calm down and think carefully. When you called the lawyer did you mention my name?"

"No, but I did say you were an accountant in one of Matt's companies."

"Well, I doubt Matt had my place bugged. I don't think he would have found out who I am just yet. He clearly had you go to the condo to control you and watch your every move. Do you remember how my uncle left me his place in Princeton?"

"Yes, I do."

"Ok, write this address down and have a cab take you there."

"I can't just leave Shamus."

"Oh, yeah, that's right. Is he coming back to the condo?"

"Yes, the investigator offered to take him back and forth."

"All right then, as soon as he gets back take a cab and pay cash. I will see you both there later. On second thought, do you think you should have that investigator check out my place?"

"Yes, that is a good idea. I have one of your keys and I will give it to him when he comes back here with Shamus."

"Ok, Dawn, I will be at my uncle's old place by 4 p.m."

Dawn starts to cry. "I can't believe this is happening."

"Dawn, honey please get a hold of yourself and do not have another drink. Pack up what you need for now and we will talk about this later. We will figure out what the best next step will be. I love you, talk to you later."

"I love you too, Ken."

After he hangs up the phone Ken goes into a rage and throws things all over his apartment.

"Damn, how did I ever let myself get into this mess! I am not a dumb man! Damn, I must have been thinking with my prick!"

Ken hastens to pack a few things and calls a cab as well. Ken arrives at 2 p.m. in Princeton and walks to town for a drink at a local pub after which he walks around the Princeton University campus, which is perhaps one of the most beautiful in the country. It is a bucolic and serene setting with mature trees and old English-style structures covered in ivy. He tries to clear his head and heads back to the house to meet Dawn and Shamus.

He sees the cab pull up and looks down the street to see if they were followed. They come to the door with some small overnight bags. Ken and her embrace while Shamus picks up the bags and brings them to the house.

"Ken, please tell me this is going to be ok."

"Darling it is too soon to say how bad this is going to get. Let's just breath and sit here for a bit."

"Shamus, would you mind giving Ken and I some privacy?"

"Honey, let's just go upstairs and talk. Shamus, here is how you operate the TV; there are lots of football games on. Dawn and I will be down soon."

"I am not a total fool, Ken, you don't have to treat me like a child."

"I am sorry, Shamus, we are just trying to piece things together and need some space."

"Ok, Ken, you guys go do that; I will be fine."

The two head upstairs and close the bedroom door. Dawn breaks down in tears and falls onto the bed, curling into a fetal position. Ken sits beside her and places his hand on her thigh and says that they should speak to the lawyer together on Monday.

"Do you have the phone number for that investigator guy?"

"Yes, his card is in my purse."

She hands him the card and calls but gets his voicemail.

"Hey, this is Ken; please let me know if you located anything in my place. Dawn and I are out of town and I can be reached at the following number."

Ken lies down next to Dawn and the two fall asleep from mental exhaustion and all the drinking. Shamus continues to watch football well into the evening.

Chapter 15:
THE WALK

Back at the Biancas', the morning sun is just making its way into the sky along the ridge of the hills outside the village. The vineyards weave their way through the hills like silky, gentle waves rolling across the landscape. Rose is the first to rise, as she always does, to take a walk at the earliest of hours. She thinks this is a good opportunity to be with her father and walks quietly to his door. Matt is fast asleep when he hears the very faint taps and whispers at the door. He tells Rose to come in. He senses that it is early and asks her if everything is all right.

"Dad, I have been walking very early every morning around the village as well as up in the vineyards. I thought you might like to do that with me. I remember we used to go on walks all the time, especially on Saturday afternoon when we would walk into town and get ice cream sundaes."

"Yes, Rose, of course I would like to join you; I used to look forward to those walks. Give me a few minutes to get changed and I will meet you downstairs, ok, honey?"

"Ok, I will be in the kitchen making us some coffee."

Constance is already busy in the kitchen, getting things ready for breakfast, and Rose greets her with a hug and kiss.

"Dad and I are going for an early walk."

"Oh, that is very nice. Would you like me to prepare a basket so that the two of you can have some snacks along the way?"

"That is a great idea, thank you very much, Aunt Constance."

Constance places bread, cheese, olives, and crackers neatly in the basket. Matt walks into the room and asks what the two of them are up to.

"Well, Matt, Rose and I thought it may be nice if I put your breakfast in a basket so the two of you could snack on your walk. I think you take your coffee black, Matt, yes?"

"Perfect, Constance, you are the best."

The coffees are poured and the two give Constance a hug and then head out on their little journey.

The village is very quiet and they are the only ones out on the street this Saturday morning.

"Rose, I loved all of our walks and I am so glad you had this idea. Hey, maybe we can find some gelato later, just like we used to do with the ice cream sundaes."

"I know a place not far from here, so let's stop there on our way back—but don't tell Constance, because she makes her own and would be upset if we bought it somewhere."

"For sure, Rose, she is very proud."

The two make their way out of the village and into the dirt service road that runs through the vineyards. It takes them nearly an hour to reach a point high on the hills that overlooks the village and it is there that they decide it's a good place to sit and take a break. They find a nice grassy patch and kick off their shoes and sit down. The coffee is still hot as Rose hands it to Matt and says, "Be careful, dad; this I still hot."

Matt puts it down on the ground next to him.

"Rose, sit back and close your eyes and smell this place."

She takes a few breaths in and says, "It is amazing

how much more we seem to smell when we shut down the other senses."

"Yes, Rose, this place is intoxicating to me. Ever since I got to the village I've been overwhelmed by the smell of Sicilian foods cooking, and the scent of the air on the drive down through the vineyards on our way in was so powerful. I could smell so many things: flowers, grapes, and a hint of baking bread. I think everyone should pause and close their eyes everyday for a little while, if not to take in the senses of the world around them but to think about things without the visual stimulation. God has blessed the blind with the ability not only to have sharpened senses but to be able to think in ways that I am not sure that sighted people can."

"Dad, I am sure you are right; that is why people are always trying to find better ways to meditate."

"Rose, honey, it has been such a long time since I have felt your face. Is it ok if I do that now, honey?"

"Of course, dad."

She pulls herself closer and takes his hands and places them on each side of her face.

"Oh my, you have the same face structure your mother did," Matt says. "Rose, you have become a beautiful young woman."

He begins to tear up, as does she. The two embrace and he says how much he misses his deceased wife and the times they had as a family.

"Dad, I have to tell you something that has been on my mind."

"Ok, sure, Rose, what?"

"Well, you have not been speaking much if at all about Dawn, and you just don't seem to be your usual self. Is something bothering you?"

"I am fine, honey. Perhaps I have been working too hard and realized that it was time to take a break, and what better place and people to do it with?"

"Yes, you certainly seemed to be working very hard lately and I think you are at the point where you can begin to delegate more and more to others."

"You are so right, Rose. Your mother was my balance and we were a great team. I was able to balance things much better when she was alive. My second marriage, on the other hand, has been different . . . Oh, damn it—I must not keep this from you. I didn't want to tell you this because I don't want you to be upset or leave your vacation, but I just cant hold it in anymore. Dawn and I are splitting up."

"Oh my God, dad, are you ok? What has happened?

What can I do? I knew something was wrong. Dad, you must know that you can always come to me with anything. I am stronger than you think."

"I am sorry, Rose, I was just being my usual, over-protective self."

"Why has this come about, dad?"

"Well, let's just say she has not been faithful."

"Oh my, I am so sorry, dad. Do you want to talk about it?"

"My dear, I walked in on something that I should not have and I suspect there is more to it."

"What do you mean?"

"Well, Shamus seemed to have known about this as well because when I got home he was on the door at that time. I came home early to surprise Dawn with some flowers. The buzzer to the apartment wasn't working and so the cheaters did not know I was coming, since Shamus's alert had not gone through. When I got out of the elevator and into the apartment they were in my bed."

Rose gasps.

"That must have been so horrible for you, dad. I am so sorry. What can I do?"

"Rose, I have an entire team taking care of this. Being here with you and the others in this place is all I need."

"I love you, dad, and will always be there for you."

"I love you too, Rose; we will all be fine. You have to promise me you will stay here and finish your studies. I am worried that you will want to leave early because of me, and all that is going on."

"Dad, you need to do something for me then. Please just call as mush as you can and tell me what is happening."

"I will, Rose, and I am sorry that I kept this from you. You know how protective I am and that often clouds my judgment. You are so much like your mother; some man will be very lucky to have you some day."

"Dad, do you remember what you told me years ago about relationships?"

"Well, I have said many things."

"Well, one of the lessons about people that you have spoken about is that there are no chance meetings in life and that each person we come into contact with, albeit sometimes for a fleeting moment, has profound meaning."

"Rose, I really believe that, and each moment is

precious in its own way. The key is to tune into that as best as you can every day."

"Well, then because of that, I am sure I will meet the right man when the time comes."

"Rose, I will also tell you of one thing about meeting people. I tend to sense the good in people and often miss signals of ill will. I think there is good in all of us, even those that have done horrible things. Since you are like your mother I think you have always been a good judge of people but have learned that if you feel something is a bit off and cannot explain it there is likely something wrong. I am often so focused on the positive aspects of people that I miss what may be underlying problems within a person. I am not speaking of the troubles of life; we all carry those burdens everyday. Rather, I am referring to those who harbor bad intentions. Frankly, after thinking back on my early days with Dawn, I had a feeling that there was something just not right with her or our relationship."

"Dad, honestly, we thought the same thing, but none of us wanted to say anything because you seemed so happy and we were thrilled to see you in love again. As I think back a bit, dad, I did notice that you seemed more serious around Dawn. Your sense of humor seemed to be dampened and you were less jovial in general. I think back to when we were kids and we played jokes on each other. Those were great times, dad. One of my favorites was on April Fools' Day, when we moved the furniture around and you had to navigate around the new obstacles and we would run and hide and you would have to find us. That was so much fun."

"Rose, that was a great time; we used to laugh all day when we did that."

"Dad, do you remember the time we went to Yankee stadium and you bumped into a stranger and knocked his soda all over him. He said, "Hey, buddy, are you blind or something?" and you replied, "Yes, as a matter of fact I am, but was fired from the Yankees for my poor fielding." You both had a laugh at that and went on your way. I bet that man still remembers that, dad. When you came riding into village driving the car the other day I saw the old Matt Messina: the man who loves to joke around. I am really sorry about what happened with Dawn, dad, but I have to tell you I already see signs of the happy man we all knew coming back to us."

"Rose from now on lets not keep anything from each other."

"Absolutely, dad, no more of that."

"Rose, your mother and I often spoke of this and

we used to speak about letting you and your brothers learn by making mistakes. I still think that is a good process but when we are aware of major life events that may cause us to make a wrong turn we should be open with each other."

"I understand, dad, and we should speak to Mark and Luke about this as well."

"Yes, lets do that. When we get back to the house we should share this discussion with everyone. For now, let's enjoy this time and place together."

The two remain on the hill for a couple of hours and make their way back down to the village. As the they make their way into the village Matt asks Rose if there are any wine stores close by, as they should pick up some nice selections since they will be visiting with people later that day.

"We should also bring some to the Chiazanos."

"Dad, there is a great place right around the corner from the Biancas'; let's go there. They also have great cheeses and lots of other things as well."

"Perfect, Rose, let's do that."

As the two enter the store the owner recognizes them from the welcoming event in the town square and the now locally famous stickball game. The owner makes his way from behind the counter and introduces himself. He thanks Matt for teaching his son a life lesson, as one of the kids on the other team was his youngest boy.

"I don't think my boy will ever underestimate someone in the future. They were certain they would win that game. I am grateful to you for what he has learned."

"You are kind; it was my pleasure to be able to play with the boys, and frankly, it helped me as well. My family is here to reconnect with our roots and to have fun like that made me feel like a kid again. It also brought back memories of when I used to play with my own children like that. We would like to purchase some wine for our hosts; do you have some nice Amarones or Barolos?"

"My friend, I have the best here. Do you have price concerns, and how many would you like?"

"No concerns on price, and we will need a case of each, please."

"Right away, sir, and may we deliver?"

"Great, yes, please bring it to the Bianca's home on Via Cucina; do you know them?"

"Yes, no problem, we will have it in fifteen minutes."

Matt pays in cash and they return home. The Bianca family is all set up and in the back courtyard awaiting their arrival. Constance offers up some fresh lemonade made from the fruit on their property.

"You two look like you had a real hike."

They are covered in dust up to the knees.

"Not only did we have a great hike, but we had time to talk about a lot of things, and Rose now knows about my situation with Dawn," Matt says. "I feel bad for having kept it from her and making you all lie to her. She will stay here and finish her studies after we return and I promised to keep her up-to-date on all that is happening."

Everyone is relieved by the news.

"So, dad, it looks like we have a busy couple of days ahead," says Luke. "Grandpa's old house this afternoon, the Chiazano's this evening, and then church and the Sunday feast. Good thing we are doing a lot of walking or else we would need to see a tailor to let our waists out."

There is a knock on the door. It is the wine delivery. Matt tells the Bianco's one case is theirs and the rest will be taken to their hosts this evening, Father Passante, and the gentleman that lives in his father's old home. Costance is thrilled, as they love the selection.

"Thanks so much, Matt."

"It is the least we could do, Constance. Alfredo will be helping with the food tomorrow, and please remember my flight crew will be joining us. They do such a great job for us and this will be a great chance for them to have a real Italian feast."

Giovanni offers to drive up to Alfredo's this afternoon to pick up everything in advance.

"That is a great idea; I will need a lot of time to organize everything," Constance says.

"Rose and I should go clean up and get ready to go," Matt says.

"Ok, dad, Luke and I are all set and ready when you are."

"You boys should bring this bottle of scotch with you," Joseph says. "It's very nice that your father is bringing Mr. De Francesco this, as he seems to live his days with a bottle of scotch in his hand. Also you all should know: Mr. De Francesco is still living in the very distant past as a pilot in the Spanish Civil War and World War II. He is a very proud man and is never politically correct, but he will love to have company and I am sure be happy to show you the home—even more so because of the bottle of scotch."

Giovanni tells them to have a great time and heads out to Alfredo's to get the provisions for tomorrow's feast. Matt and Rose are ready in quick order and come back down to join the men.

"I am always impressed by how quick Rose gets ready," Joseph says. "Most women take hours."

"Yes, my Rose was like that, and I think my daughter picked up that trait."

Constance is quick to add that natural beauty does not require much, if anything, to prepare to go out.

"Oh, thank you so much, Aunt, I love you."

Constance gives her a hug and kiss and asks if they will be back after the visit to the old family home before they go to the Chiazano's.

"Don't plan on us coming back, Aunt; I think we will have very little time and I want to show the boys and dad around the village a bit."

"Very well, you all have a wonderful day."

As they leave the home the village is bustling with activity and the day is beautiful with bright sun and warm air. Matt smells the scent of lemon gelato in the air as they pass a street vendor.

"Kids, is that lemon gelato I smell?"

"You bet it is, dad," Mark says.

"We all have to get some. Mark, here is some money; give the man a big tip and get us some, please."

"Oh my gosh!" Luke exclaims. "This is the best ice cream—or whatever this is—that I have ever had."

They all take a moment to sit on a bench and enjoy it.

"You know, this would be a big hit at home, dad," says Mark. "We could franchise this stuff and make a killing at the Jersey Shore and other beach places."

"Hey, son, you know, I think you are on to something. There is a huge profit margin in this stuff."

"Oh boy, we Messinas never stop, do we?" Rose says.

Luke laughs.

"No, we can't; you are right. I think its in our blood to see these kinds of things."

"Well, I do like to be diversified, and you can scale up something like this without whole bunch of overhead," adds Matt. "Mark, why don't you write up the business plan and show it to some of the accountants as well as planners in our Madison Avenue office when we get back. If nothing else, this will be a good lesson for you kids. Your uncle and I started with nothing and built a nice business. If it were not for that we would not have had the fundamentals in place when mom and I came into all that money from the Alucia. It's painful to see how some of the other investors, who made tens of millions from that, lost some—if not all—of their money with crazy schemes. Your uncle and I knew our business inside and out and learned how to grow profitably with little or no debt. When our windfall came in from the treasure we used the money wisely to scale up the business by purchasing our competitors for cash and keeping the key owners and employees involved with their clients. We offered them a very nice package that would lift the burdens of owning a business while making them more money by providing access to better markets and infrastructure.

"We made only one bad acquisition and it really wasn't bad from a financial perspective but rather the attitude of the former owner, who was hard to work with. He had seller's remorse and was critical of our processes and thus did not play well in the sandbox with others in the company. In fact, many of the clients he had cancelled because of this.

"We learned a big lesson from that. It is not only the financial arrangements that need to be examined and thought through very well but also the personalities and intentions of the people you will be working with. I know that one seems a bit obvious, and that it may be overcome with the right management, but no deal is worth it if it disrupts your overall business. There are too many good opportunities out there to be dealing with difficult people.

"Remember kids, you can't do a good deal with a bad person, no matter how hard you try. By the time you kids were born, your uncle and I had purchased over forty insurance agencies and brokers in the New York and New Jersey area. Most of those folks, as you know, have their kids working in the combined company today. Your most important asset is your people. It is amazing to me how many companies treat their employees like chess pawns and don't care about things like balancing work with family. I used to call your mother our 'Chief People Officer.' She was the one that made sure all of our employees were happy with their positions and made sure that they had the proper access to resources and time away from the office to take care of their families.

"In fact, when we were a smaller company we would pay for the trip expenses of anyone who wanted to go to the Holy Land or to places that made them connect with their religious core. I forgot about that until I just brought it up, and you know something, I think I will speak to your uncle and see if he wants to reinstate that for our people. We have grown so large and so fast I lost track of things like that. Once we had the forty or so firms under our roof, that provided the critical mass to make one of the largest privately held insurance distribution companies in the northeast. The model for the firms we bought became turnkey for them and us. We really had the acquisition process down to a neat and tidy offering, and it was then that we went viral. The folks we bought had better lives after they sold and we started to get a great reputation for doing the right thing. Companies were coming to us; we often did not have to go to them.

"I recall one closing we had. After we got back to the office, your uncle and I got a call from the gentleman who had just sold us his business. He pointed out that the lawyers on both sides had messed up and due to the date of the closing he lost $137,000 in commissions payable to him. After thinking about it for a brief moment we knew he was right. He asked that we draw up a document to get back to the lawyers. We said no worry, there's no need for that. Just come by and pick up a check right now. He was moved by this , and the word of things like that got out.

It's amazing what a person or company can do if they do the right thing. It's simple when you think about it. We hear it in church and other places all the time but we forget and have to remain vigilant every day."

Matt starts to get a bit agitated and clenches the hand not holding the gelato.

"Yes, vigilant—keep your eyes open and never let your guard down."

"Dad, are you ok?" Rose asks.

"Oh, I am sorry; I was just thinking about Dawn. You know, what I meant to say was keep your heart and eyes open kids. The test for me now is to keep my heart open in the face of this issue we now face with Dawn. Let's not let her drag us down in any way. We are Messinas and we must remain positive and directed in our thoughts and actions. This is what your mother would have wanted as well."

Chapter 16:
HIGH ANXIETY

Back in Princeton, Dawn and Ken are just waking up. Shamus is already playing video games downstairs. As Dawn opens her eyes she starts to cry and embraces Ken.

"Please tell me you will never leave me, Ken, no matter what Matt and his people try to do to us."

"I promise, Dawn; I love you more than I have ever loved anyone in my life and I knew it the moment I saw you. I felt so conflicted. Matt was always so good to me and all of us at the office, but the emotions overtook me. It was as though you jumped right into my soul; all my thoughts were of you."

"Ken, it was the same for me, and I am so worried that Matt will find out about the money and we will be sent to prison for a long time. Maybe we should leave the country now, before all this hits the fan."

"Dawn, you know, I was thinking about doing that

as well while I was on my way here yesterday. I took a long walk around the Princeton University Campus; it is such a beautiful place and it has always had a way of calming me. When I was a little boy I loved staying here on weekends with my cousin. We played great hide and seek games all around these stately old buildings and beautiful grounds. As we grew older we gravitated to the donut shops, then pizza places, and ultimately the established old pubs. It is an amazing place to raise a child. We didn't have a lot, but we were a close family. We made our own fun. Today, kids of all ages, like Shamus down there on his video game, miss out on the basics of life. They are too hung up with their electronics.

"All sorts of things were running through my head as I walked yesterday. If I were to suddenly disappear from work, I am sure the auditors will be called in. It will take the forensic financial people a long time to detail the trail of the money, but ultimately it may not get to the end place it resides in Switzerland. So, I need to initiate this process by giving notice that I will be leaving for another position with new company. It will give me time to cross check all that I have done in the system and to create more confusion within the files. If I simply say that I am going to take another job it would not raise any concerns initially. If a sharp financial officer were to take my position, he or she may pick up on some things over time, but by then we would be long gone.

"I think you should meet as early as possible on Monday with your lawyer. I am concerned that device may have recorded you and that may get things moving fast. We need to find that out first, before we make any moves."

"Ken, I am sure all she can tell us on that front is whether the listening thing was working or not and we will not know what was picked up on the other end."

"Yes, you are right; our minds are racing all over the place and we need to calm down and think rationally."

"Look, the criminal stuff is all on me, Dawn. I think you will be in the clear. My guess is that Matt will try to play hardball with you to get to me and retrace the company money."

"Yes, he would, and I am a conspirator, Ken, so I think I will have criminal exposure as well. Maybe we should just make a run for it? Do you have access to the money if we just get to Switzerland?"

"Yes, I could literally walk in and get it."

"Well, Ken, I don't have to think too hard and long about this. My vote is to get over there as soon as possible and erase our trail from there. But what am I to do with poor Shamus?"

Ken stares at her for a moment and lowers his head then looks back up at her.

"Honey, he is your brother. He has to come with us."

"Oh my, Ken, I don't think that is a great idea. I love him, but he will be the cause of our downfall. Look at what has happened already. I have an aunt that lives down in Palm Beach, Florida. He loves her and always likes to visit with her. She loves him and always wanted him to live with her. When her husband died five years ago Shamus went to stay with her and they kept each other company. I really think that would be best. We could tell him that you and I will be working hard to make plans up here and that he could take a vacation with my aunt. Let's go tell him that now."

"All right, Dawn; that sounds fine."

The two get dressed and head downstairs to make some breakfast and call Shamus in to join them.

"Shamus, Ken and I were thinking it would be good if you visit with Aunt Pauline down in Palm Beach. She is so lonely, and this may be a good time for you to get away from all this while Ken and I work on getting things cleared up."

Shamus stands up and walks over to the kitchen window to look out. Without turning around he says: "I may be a bit slow but I am not stupid. I know you guys are up to something and that you are mad at me. I did not cause all these troubles, you guys did. I will go to Aunt Pauline's and that will be fine with me. She loves me and I love her and I want to be away from all of this scary stuff."

Dawn gets up from her chair and walks over to put her arms around Shamus. "I love you, Shamus, and we are not mad at you. We really are not. Ken and I have so much to do to get us out of this mess. You are right, and this is not all your fault. Something was bound to happen, and who knew that it was going to be a silly broken buzzer at the Park Avenue place that would expose us? We put you in the middle of all this and should have never done that."

She continues to hug him and begins to cry. Through all of this Shamus does not turn around to look at her. He gently pushes her away and asks to be alone for a while.

"You can send me to Aunt Pauline's whenever you want, but please do it soon. I want to be away from all of this. It upsets me and makes me shake."

"Shamus, I will call her this afternoon and make the arrangements."

"Please Dawn, get me out of here as fast as you can."

"I will call her right now, but, Shamus, you must not breath one word of this to anyone ever again. Do you understand me?"

"Dawn, I don't want to talk about it anyway; it only upsets me."

Ken tells Shamus that he, too, is sorry for having gotten him in trouble and vows that he will be ok and will not have to worry about this anymore.

"Ken, you may think I am stupid but I know what the word conspiracy means and there is nothing you can do to completely protect me, but thank you for offering. I am done talking about this; Dawn please call her now and let me listen to what you tell her."

Dawn walks over to the wall phone in the kitchen and calls Aunt Pauline. At the other end of the phone is a beautiful woman in her late fifties. Aunt Pauline is a true belle of Palm Beach. She is the president of the the Woman's Club of Palm Beach and a member of the most prestigious country club in the area. She is an elegant, shapely woman who never remarried after her husband's death many years prior.

"Dawn, my dear, it's good to hear your voice; it has been quite some time since we spoke. How are you?"

"Well, Aunt Pauline, frankly things are a bit stressed up here. Matt and I are having some difficulties in our marriage and I need some time to focus on that. I have very little time right now to spend with Shamus as I typically would on the weekends but feel bad that he will be alone. Would you mind if Shamus visited for a while?"

"Oh my, I am so sorry to hear about this; you must do all you can to repair whatever it going wrong with your relationship. A marriage is a sacred thing and you two must do all you can to stay together. I am shocked by this, Dawn, and will not ask you any details unless you seek my guidance but pray it all works out for you and Matt. I love Shamus and he is welcome to stay here with me as long as he wants. I am very busy with social activities here, but Shamus and I have always had a special bond. I am not sure he has ever told you, but one of his favorite things is going to church with me on Sunday. Church here is more than a service but rather a social gathering of nice people and lots of activities. We always go to lunch at the Breakers right after service with a group of friends, after which Shamus and I walk on the beach. It is my favorite day of the week and I am sure we will do our usual routine. Sorry, honey, I am rambling a bit, but I am so excited that he will be coming and yet I am sad for you and Matt. I promise I will keep you in my prayer concerns daily. When will Shamus be coming?"

"I will be making the flight arrangements today and will call you back, but Shamus would like to come soon, so I think within the next couple of days; would that be ok?"

"Certainly, darling, please call me as soon as you know. Please take care of yourself, Dawn; I am sure everything will work out."

"I love you, Aunt APauline, speak to you very soon."

"Love you, too, and send a kiss to Shamus for me and tell him I am very excited that will be coming to be with me."

"I will, Aunt Pauline, take care and thanks so much."

As she hangs up the phone Shamus says he doesn't like lying to people and goes on to say that the first time he did it in his life was this thing with Matt and now this.

"Shamus, I did not lie; I told her Matt and I are having trouble. What I said was true, just not all the details."

As Shamus walks away he says: "Call it what you will."

Ken is shaking his head in frustration.

"This is all so screwed up, but I guess it's our best option. We should not bring Shamus back to his place to get his stuff. I am sure it is being watched."

"I have a friend that helps me with chores and errands that Matt never knew about," Dawn says. "I will have her go to his place and pack some things and drop it off at the Jersey condo. I am sure that will not look suspicious. I would think that they would expect me and Shamus to be spending time together."

Ken suggests that Dawn should call the investigator to see if there were any surveillance bugs found in Shamus's place first.

"That's a good idea; I will call him now."

The investigator tells her that another device was found in Shamus's house phone.

"Your husband's people must mobilize very quickly; you had better be very careful what you say and who you say it to."

"Thank you, I will be speaking to Jeff shortly."

She puts the phone down and sits at the kitchen table, puts her head in her hands, and begins to sob.

"What is it Dawn? What now?"

"He found another device in Shamus's apartment."

"Where was it?"

"It was in the phone."

"I think we should ask your brother to think about any conversations he has had in the last couple of days on his phone."

"I don't want to alarm him any more. I think we should just get the hell out of town as fast as we can. I will have Shamus's things brought to the condo this afternoon and I can pick them up on my way to the airport. I will get him on the evening flight to Palm Beach at Newark and come back here after. If I am followed by anyone to the airport they will only see Shamus leaving, but I will be watching to see if we are being followed."

"Dawn, I think you are right. If they went through all this trouble to bug places that fast they will prosecute the living daylights out of us when they find out about the money and I bet a criminal act will void your prenuptial agreement. We should get out while the going is good. You go ahead and make that flight for Shamus and contact your aunt. I will go upstairs and start working on international flights from the other line. Uncle still has a separate phone line up there from when he used to work at home."

Dawn is able to secure a direct flight at 8:30 p.m. and contacts her friend with a list of things to bring from Shamus's home. Fortunately, she still had keys for both places. Ken secures tickets to Switzerland that leave from JFK the next evening. Dawn heads back upstairs where the two plan out details.

"Dawn, I don't think it is a good idea for me to go back to my place. I have enough with me here and only need a few other things that I can pick up in town."

"All right, I need some things from my place in the Jersey condo but it will look like only Shamus is leaving to go somewhere if I am tailed to the airport and only he gets out of the car."

Shamus is told to be ready in a half hour and Ken calls them a cab.

"Wow, you two move fast, don't ya? Are you going to tell me what you are going to do about all of this while I am with Aunt Pauline?"

"Ken and I will be working with Jeff to make sure

all of this will be settled in a proper way with Matt. We don't want this to drag on and we want you to be happy too, Shamus."

"That is a good idea, Dawn. Matt is a good man and he and his family do not deserve any of this. His heart is broken, but I think he is a forgiving person."

"Yes, Shamus, he is, and for that reason I know we will all be ok."

Dawn cannot help but think deeply about what Shamus has just said.

"Ken and I will be upstairs working on some things, honey. Pack up what you brought and we will have your things put together for you from your place. Bonnie is packing a couple of bags for you and meeting us at the Jersey condo."

Ken and Dawn chat for a few minutes while waiting for the cab.

"Ken, I think Shamus is right; Matt is a forgiving person."

"Dawn, stop and think. We committed a crime and you cheated on him. Forgiving or not, this is beyond anyone's capacity to forgive and we cannot take that chance. He also has business partners that will weigh in on this, along with the legal authorities. We are moving too fast because we think we have to and when people move fast they make mistakes. Get back here as soon as you drop off Shamus."

"Ken, maybe I should stop at the airport hotel for a snack or something and have the cab wait for me; this will give me a chance to see if I am being followed."

"Wow, we have become so paranoid that we think of things like that, Dawn. Yes, I guess you are right; that is a good precaution."

The cab's horn is sounding outside and Shamus yells up to her to get going. The two hug and Ken says, "Please keep a cool head; we are going to be just fine. Soon we will be far from all of this and together forever."

The two exchange a long hug and kiss. Dawn and Shamus leave in the cab and Ken's mind begins to race. He is thinking that he should make a break for it and leave Dawn behind, and he begins to pace back and forth. As soon as the cab is out of sight Ken puts on his jacket and once again takes a long walk around the Princeton University campus. His mind is racing but this place calms him. He thinks that if he leaves Dawn will throw him under the bus and turn on him to make a deal with Matt. Then he thinks he is better off without her and more likely to get away and start a new life.

Ken's mind races and his thoughts become reckless. He gathers himself, takes a deep breath, and focuses his attention on the Princeton campus. The trees on the grounds are majestic and shade the buildings, each of which is a masterpiece in its own special way. They range from Collegiate Gothic to modern. Sculpture and artwork of all kinds adorn the well-groomed walkways. It is a very peaceful place. The chapel at Princeton is nothing short of breathtaking. Its doors are open and Ken decides to walk in. You can almost smell the history inside it. Someone famous has literally sat in every pew in the sanctuary. He decides to sit for a while and gather his thoughts. Although raised in the Christian faith, he has not been to church nor lived a Christian lifestyle in many years. He is not feeling very connected with God, and now takes this moment to pray and ask for guidance.

"God I know you are not happy with me and I haven't lived a life that was the way I was taught. Temptation and greed have gotten the best of me and I don't know what to do. Please help guide me."

He sits for a few moments and leaves the chapel to walk to Lake Carnegie, where the boathouse is active and the Rowing Team is actively practicing their techniques. It is an awesome sight to see so many men working in complete synchronous precision to propel their crafts through the water at high speed. He sits there for what seems to be hours and heads over to his favorite pubs in town where he proceeds to drink himself into oblivion. He is a very good-looking man and several young coeds find him attractive. He starts to flirt with a few but is not too drunk to realize that he must start to walk home before things take another direction. He buys a round of drinks for the students at the bar and starts to walk home. Ken notices that the doors to the chapel are still open and decides to cross the street and visit once again.

This time he walks all the way to the front pew, passing just a few scattered people sitting quietly. He makes eye contact with a few. One person notices he is wobbling and likely drunk. Ken stables himself and sits in the first pew directly in front of the altar. The person who had seen him walks up to the pew behind him and sits quietly. He is a very old man who is dressed like someone who has been homeless on the streets. He leans forward and whispers to Ken.

"I saw you come in earlier, and you decided to come back. The Lord is happy that you did. Keep your mind open and listen for his words."

Ken listens to his every word but does not turn

around. He stays still for several minutes, staring at all the images in the stained glass. After some time of reflection he gets up and walks out of the church, but not a soul is left there but him. He is greeted by a maintenance worker on the steps who wishes him a good day, which Ken reciprocates.

Meanwhile Dawn has already picked up Shamus's things at the condo and they are heading to Newark Airport. Shamus keeps asking why she has been looking out the back window so much. She thinks quick and says that she was concerned that since she told the taxi driver to go fast that a police car may stop them.

"Don't worry, Dawn, that would not be your ticket; it would be his."

"Of course, Shamus, I guess I would just feel bad."

"How long do you think I will be at Aunt Pauline's, Dawn?"

"I am not sure, Shamus. It may take Ken and I a while to go over all these things and we want to do it in a matter that does not upset Matt any more than he already is."

Shamus has no idea that he may never see his sister again.

"Dawn, I am very scared, so please make sure you

and Ken do the right thing and just come clean and ask Matt to forgive you."

"Shamus, yes, that is the best thing, honey."

All the time she is thinking about how her horrible lies are piling up on top of each other. Soon they are at the airport terminal and Dawn tells the cab driver to take out all the luggage except the two Luis Viton bags. She and Shamus embrace at the curb and she tells him everything will be all right and that Aunt Pauline will take very good care of him, to which Shamus replies: "Dawn, Aunt Pauline and I take good care of each other."

"You are so right, Shamus. Please be sure and call me when you arrive. Here is the phone number at Ken's place in Princeton."

"Dawn, why are you bringing those two large pieces of luggage to Ken's place?"

"I just don't know how many days we will be there, and you know me: I can never decide what to wear."

They embrace again and the two bid each other a tearful farewell.

Dawn watches Shamus from inside the cab until he is well inside the terminal building and while doing so she scans the area behind the cab to see if they are being followed; sure enough, a black Suburban appears to be doing just that.

"Driver, please stop at the airport hotel over in the parking area. I need to stop and use the restroom."

"Yes, mam, I will be right out front waiting for you."

She walks into the lobby and down a hall that still provides a view to the front so she can still see the cab and the black Suburban, which was parked several cars away. She comes up with a plan to get her bags brought into the hotel to give the appearance of checking in. She calls Ken, who by this time has made it home.

"Ken, I was definitely followed. I am thinking the best thing is to make it look like I am checking in here and switch to a different car in a little while."

"When did you notice that you were being followed?"

"I think I saw a black Suburban on the highway but did not think anything of it until I saw it pull in several cars behind us and no one got in or out. Then, as we pulled away, it did as well and now it is in front of the hotel and no one has exited or approached the vehicle except the doorman."

"Dawn, you are right. Go have your bags brought in and see if the vehicle leaves."

"I will, Ken; stay by the phone and wait for my call."

Dawn has her bags brought in and placed near her table at the restaurant, which still has a good view of the Suburban. She orders a Martini and slowly sips it, all the time watching the vehicle outside. She is startled by a well-dressed man offering to buy her drink. She thanks the man but says no.

"Well, sure, but a pretty lady should never drink alone," he says, and walks away.

She finishes the drink and orders another and as the second drink comes the Suburban pulls away and exits the main gate to the airport exit. She quickly downs that Martini and leaves some money on the table and asks the doorman to hail her a cab. It comes quickly as they are lined up for all the travelers. She is soon on her way but realizes she forgot to call Ken. She asks the cab driver to pull over at the first rest stop and she makes the call from there.

"Ken, I shook him and I am on the way home. I am sure whoever was watching thinks I am leaving from Newark and certainly saw Shamus leave from there. Our tickets are out of JFK tomorrow, right?"

"Yes, we leave at 5 p.m."

"Great, they will be confused; Ken, see you in about a half hour or so. I love you."

"I love you too, Dawn."

The cab arrives at the house and the driver assists with her things. As her bags are brought into the house she can see Ken is noticeably intoxicated and sits on the couch next to him and gives him a long hug.

"Did you make sure no one followed you the rest of the way?"

"Yes, no one did. I was watching the entire way and I am sure the cab driver noticed that. Do we have any coffee in your uncle's kitchen? I think we both need it."

"Yes, it's in the cabinet to the right of the sink. Thanks, I can barely move right now."

Dawn comes back in with the coffee and Ken is staring out the window and is without words. She tells him that he looks like she feels.

"Ken, I can't believe what has happened, and now I shipped my brother away and I am thinking I may never see him again. We are awful people. How could we have done this?"

"Dawn, we let booze cloud our judgment and things got away from us. We let booze drive all of our thoughts, and when you look at it we never did anything together without being drunk. Greed, sex, and booze took over our way of life."

"Yes, you know, we are bad people, but we still love each other and can do good things. Ken, Shamus still thinks we she should ask Matt to forgive us; maybe if we tell him about the money and return it he will not press charges."

"Dawn, I have been thinking about that all day, and you know I even went to church and sat there and tried to see if an answer to this would come into my head."

"Wow, drunk at church?"

"Yes, what a mess. Well, after all that praying and thinking I think we should make a run for it tomorrow. Although Matt is the principle owner of the business, he does have partners and his kids will drill us into the ground. Never forget the bond he has with those kids. That is the emotion that will drive his thinking and for that reason it is clear in my mind we have to get on that plane tomorrow."

"You are so right, Ken; I am sure those kids never really liked me and they never got over the death of their mother. Yes, lets make a run for it and never look back. How much do you think will be in the account in Switzerland?"

"At last check it was a bit over ten million."

"Well, if we are smart about it we can live the rest of our lives on that, Ken. We both came from nothing and that is a lot to us—and a drop in the bucket to Matt."

"Dawn, I was once told that criminals go through a mental process of justifying what they stole. Let's call it what it is: we are thieves."

"I know that, Ken, I was just trying to say that it is not a lot of money and that we will have to live smart once we get there. It may be a lot of money to most people but we are rather young and will have to be very careful."

"Oh for God's sake, Dawn, you sound like a financial planning commercial."

Ken storms out of the room and Dawn follows him into the bedroom.

"Ken, I am sorry we are both stressed, and I guess I am making unthoughtful small talk. We have never fought, and please, let's not do this now."

"Dawn, you have no family other than Shamus, and you really were not that close. I have a huge family and I am close to all of them. I cannot imagine what this will do to my parents. This will break their hearts, to say the least."

"Ken, let's juts get on that plane tomorrow and figure things out over there. I am sure you will be able to see your family once we get things in place over there."

"Dawn, we will be fugitives and people will track my family as well. This is a disaster. Maybe if we turn ourselves in the system will be easier on us."

"Ken, come lay down; let me give you a back rub so you can clear your head."

Dawn takes off Ken's clothes and she removes hers as well. She knows how to put Ken in a spell. She massages every inch of his body as he falls asleep. She then lies next to him and drifts asleep as well.

Chapter 17:

The Box

Back in the village, the Messinas leave the Bianca residence and are en route to the home of Matt's parents, which is now occupied by Major De Francesco. Matt is holding the gift bottle of scotch and the family is greeted along the way by several residents of the village who recognize them. Salvator is out running errands for their visit later that evening and runs into the family.

"Hello, sir, my mother is preparing a beautiful meal for you and your family. I am out picking up some things for her. We are very much looking forward to later. You will all like Father Passante as well."

Matt tells him they are all looking forward to it and are on the way to see the home that his parents once had.

"That is nice, sir. I hope you all have a nice afternoon and will see you all later."

"Thanks, Salvatore, please tell you mother we are excited."

As Salvatore walks away he turns and says, "Yes, sir, I will."

Rose comments on how polite the young man is and goes on to say it is so sad to see them alone like that.

"Salvatore's mother is such a kind and beautiful woman," she says."

"Yes, she is quite striking," Mark adds. "She reminds me a bit of Sophia Loren."

Luke is quick to agree. Matt has a vision in his own mind.y

"You three have been painting quite a picture; the men in this village must have thoughts about her," Matt says.

The family is now turning down the street where the old family residence is.

"That must be it," Rose says, "the one with the two large, blue flower pots on each side of the door. Constance told me this before we left, dad."

"That is amazing, honey, those were there the last time I went there with your mother; things don't change much around here."

They pull back the large brass door knocker and

announce their arrival with several small bangs on the knocker plate.

The doors open and inside stands a very old man dressed in his Italian Air Force uniform. He is standing full at attention.

"Please come in; this house is yours for the day," he says.

Mark and Luke give each other a look like this old man must have lost his mind in the war.

"You think he is shell shocked or just an old delusional man?" Luke whispers to Mark.

"Luke, he may be both, and the booze has added to his colorful persona, from what I have heard."

Mr. De Francesco leads them to the living room where he has prepared a small cheese plate with olives and bread. Matt presents him with the bottle and introduces the three children.

"Thank you, sir, I will go put this in the kitchen and get us some glasses."

"Rose, please go help the Major."

"Yes, of course, dad."

As soon as the two have left the room Matt whispers to the boys.

"Hey, I heard you two whispering when we came in. Let's have a little respect for this gentleman."

"Dad, he is in his old Air Force uniform."

"I am not surprised; Mr. De Francesco was a great hero back in the day. He is a very proud man. After his wife died, all he had left were memories and his bottles."

As Rose and Mr. De Francesco come back to the room he asks them to relax and make themselves comfortable, telling them he would be will be right back.

"Does he expect us to drink that scotch with him?" Mark asks.

"Sit tight, I will get us all waters so that if we have a toast we can wash it down quick," Rose says. "I don't think we should insult him; he is very excited to have us."

She gets the waters quickly and returns just as Major De Francesco comes back into the living room, this time holding a small wooden box.

"Sir, I have something very special for you."

He places the box in Matt's hands.

"What is this, Major?"

"Well, I was having some repairs done not long ago

that required me to replace some of my stairs. It was then that the workers found this box hidden beneath one of the steps. I knew that your daughter was here in the village and was going to bring it to her, but it is my honor to present it to you today. This belonged to your parents and for some reason they hid this and may have forgotten it."

Mr. De Francesco stands at full attention and salutes Matt as he presents the box to him. The kids are astonished and Matt is a bit overwhelmed and thanks him with sincere gratitude.

"May I open it, Major?"

"Yes, sir, if you please."

"Kids, come closer; why don't you open it and tell me what is inside."

First, there is a fine cameo locket with two pictures of Matt's parents. One is at their wedding and the other is the two under a tree in their Sunday clothes. Rose places it in Matt's hands.

"My mother had been looking for this for years."

Matt gently runs his fingers over the fine handmade piece. This is a very tender moment for the family.

"There is more, dad," Rose says.

There is another small box inside the box and within it is a small book of prayers. Inside the book are two folded letters. Major De Francesco says he has not read the notes as they must certainly be private.

"Dad, may I read them?" asks Rose. "I think my Italian is good enough."

"Sure, please do, Rose."

She looks over the first note and breaks down in tears.

"Father, I don't think I can do this right now."

"What is wrong, honey?"

"Nothing, Father, this is the most beautiful love note I have ever read in my life."

Matt says that it is probably best to fold the letter back up and put it in the prayer book.

"These are my parent's most treasured things, Major. You are a true hero in every sense of the word. We cannot thank you enough for having recovered and safely returned these family treasures to us."

He stands at attention and gives another salute to the family.

"I am just doing my duty, sir. Let's toast to this occasion. Allow me to pour us all some of this fine scotch."

The major pours about a shot size for the Messinas and a full glass for himself. Matt realizes it from the length of his last pour and the kids are trying are trying not to laugh. Had it not been for the tender moment they just had they may have bust out in laughter. The major toasts: "To the health and happiness of the Messina family." The Messinas take a small sip followed by lots of water while the Major drinks nearly the whole glass.

"Mr. Messina, are you parents still with us on this earth?"

"Major, Papa died many years ago and Mama did as well shortly after. I think she died of a broken heart after Papa died. They went through so much together and never left each other's side. It was a great honor and privilege to have had parents like them. They taught me and my brother and sisters countless lessons in life and shaped us into solid citizens."

"My father and mother died when I was very young," the major says. "There was controversy surrounding their death. I was told that they died in a fire in their tailor shop were they worked every day except Sunday. Many think they were on the wrong side of the political wave that was starting to take shape in our land as we got closer to World War II. Matt, your parents were like so many who saw this terrible war coming and left for America. I hold no malice to those who left Italy or Sicily in search of a new life, but as you can see, I am proud of my service and always adorn my uniform for special occasions such as this."

Matt says they are honored that he considers their visit a special event.

"Would it be all right if my children saw the entire home, Major?"

"Yes, indeed, sir, please follow me."

The major shows them every room in the modest structure; the last of it was a small courtyard in the back of the residence.

"My Papa told many stories of how Mama and he would sit out here in the evenings and sip wine and talk for hours. They also had a fig tree and lemon tree that they cared for."

The major suggests they should have another toast. Before they would be subjected to more scotch, Rose thanks him for his kindness stating that they had to be at another engagement shortly. Matt, still clinging to the little wooden box, hands it to Rose and extends his hand to shake the major's.

"I am very grateful for you having allowed us into

your home. It was important that my children connect with their roots and experience the home of their grandparents. I am eternally thankful for this box and the special things inside. Major, you are a true hero and gentleman. We look forward to seeing you tomorrow at the feast the Biancas are preparing after church."

With that, the major stands at attention and gives the Messina family a salute and walks them to the door and bids them farewell until tomorrow.

As they hear the door close behind them, Luke turns to make sure the major is inside and says, "That felt like something out of a movie scene." Matt opens the face to the braille watch on his hand to feel the time.

"It is nearly four thirty; let's stop back at the Biancas and pick up some of the wine we purchased to bring to Salvatore's mother for dinner and Salvatore must be waiting on our arrival to help us find their home," he says.

As they get close, Mark offers to run in and get them.

"How many, dad?"

"Get two bottles of Amarone and Borolo."

"Ok, be right out."

Mark wastes no time and is back out in moments. He says there is a lot going on in there, with the Biancas busy getting ready for the feast tomorrow.

"I feel bad they are doing everything," Luke says.

"Not to worry," Matt says. "This is what makes them happy. They look forward to this every Sunday; this time it will be just a bit bigger."

"You are so right, dad," Rose adds. "I have seen this repeated many times since I have lived here."

They pass through the town square, where the same boys from the other game were playing another stick ball game. They all notice them and request a re-match. Matt says they would love to but have to be somewhere for dinner.

"I am sure we will another time," he says.

Soon they are at the door of the Chiazanos.

Mrs. Chiazano takes off her apron in the kitchen and comes to greet them at the door.

"It is so good to have you; please, come in and make yourselves comfortable."

Matt reintroduces his children.

"It is very nice to have you all; Father Passante should be here shortly."

"We brought these bottles for you and the good Father," Matt says.

"Salvatore, please take them into the pantry and serve the Messinas while I finish up in the kitchen."

"Please, let me help you get the dinner ready," Rose says.

"Thank you, Rose; that would be nice. We girls can get to know each other a bit while all the boys talk out there."

Salvatore is in the pantry pouring the wine and his mother says he may have some today, for the special occasion, as well. The home is very modest and very clean. There is barely enough room for everyone to sit in the living room but it is very cozy. As Salvator brings the wine in there is a knock on the door. It is Father Passante. All the men rise to greet him. Salvatore takes his cape-like jacket and hangs it up as though it were ritual. Sophia and Rose come out to greet him as well and everyone is introduced. There is a bit of tension in the room, as Father Passante seems quite formal, but no one expects his sense of humor to be so keen.

"Matt, I hear we have something in common," he says. "I have only one eye, but I hear from many in the village that you are a better driver than me."

There is a great deal of laughter from all in the room and the mood changes quickly and a relaxed conversation ensues. Father Passante has a gift of being able to connect with people, as does Matt, so an immediate bond is formed by the two men.

"You have a beautiful family, Matt; God has blessed you in many ways."

"Yes, indeed, Father, he has. I am fortunate in more ways than a man should be."

"I am told your first wife is deceased and you are remarried. Is your wife with you on this trip?"

Matt has never lied to a priest, and being of profoundly devout Roman Catholic faith cannot do so.

"Father, I don't want the entire village to know our personal matters, but my wife and I are having some difficulty at the moment," Matt replies in a whisper.

Salvatore and his mother are outside of the room, along with Rose, preparing the dinner; Sophia overhears these comments but does not let on to anyone that she knows.

"Well, my son, I am here in any way I can help you and your family get through this. Are you active in a church back home?"

"Yes, we are, but I must be honest: after my dear

wife, Rose, died we slowly stopped going every week. It got to the point where we would go to one service each month or so. When I was remarried my new wife was not as devout as we were and so my attendance became even less frequent, as did the children's. I am rather ashamed about that, Father."

"Matt, the good Lord would like to have your family in his house each week, but if not he only wants you and your family to be true children of God in your daily lives."

"Father, I am pleased to say that I am sure that although we are perfect we strive to be the best possible Christians we can be. My dear, departed wife and I raised the children with strong values and I am very proud of them."

"Matt, I have heard many stories of your generosity and humility. Being humble with so much bestowed upon your family is perhaps the greatest gift of all, and that is one of the many reasons I was looking forward to meeting you—not to mention wanting to recruit you and your family into the church baseball league."

Father Passante has once again turned a solemn discussion into a light moment. Rose, Sophia, and Salvatore walk in the room as the funny remark is made. The Messinas have never met a Roman Catholic priest with such a keen sense of humor and they are delighted to be with him.

"Well, I see you are enjoying each other in here," Sophia says. "Father brings great joy to Salvatore and I. The people in this village are very lucky to have him."

"Thank you, Sophia, but I feel I am the lucky one. God has brought me many blessings and one of the greatest is being able to eat here every Saturday. I think I could live on the smell of your cooking alone."

"Oh yes, indeed, Sophia. I must tell you, I feel as though I am back in my mother's home. Nothing in the United States comes close to the smell and taste of the true Sicilian meal. My first wife spent countless hours in the kitchen learning the secret family recipe with my mother and aunts. The poor thing went to Sicilian boot camp, it seemed. My mother thought I would starve unless everything was prepared the same way she had done it. When I was remarried, my second wife was more interested in making reservations than making a meal," he says laughingly.

"Well, sir, the way to a man's soul is in through food and good cooking," Sophia says. "Perhaps your wife may come to my kitchen for food boot camp someday?"

No one knows that Sophia heard the trouble with

Matt's marriage and is quick to start laying some groundwork to a relationship.

"Well, Sophia, she may just have to do that, but I am not sure if she knows the difference between a pot or pan—and please, call me Matt, not sir."

Those in the room can sense that things are taking an interesting direction with Matt and Sophia.

"Thank you, Matt; it is time we all take our places at the table. Please allow me to take you to your seat, Matt."

She places his hand on a chair at the end of the table and asks Father Passante to please be seated at his usual seat. The two men are facing each other and the others are seated at either side of the table with Rose and Sophia to Matt's left and right. All at the table join hands and bow their heads for a short prayer, which is delivered by Father Passante.

"As you can see, I am a man of few words when there is such good food waiting."

The others don't notice that Matt holds Spohia's hand a bit longer than the rest at the table after the prayer is done. The discussion around the table flows beautifully and the feeling among them is amazing. It is as though they are one family. As the wine is consumed the conversation becomes even more relaxed. Matt comments that he is the luckiest at the table to have two beautiful women on either side of him. Rose kisses her father on the right cheek and Sophia places her hand on top of Matt's left hand briefly is a gesture of thanks.

Salvatore is watching most keenly at this. His heart is warmed by his mother's smile. He has never seen her face light up like this and doesn't know what to make of it, but it makes him happy. Matt, who is having more wine than he probably should, says, "I want you all to come to the United States as our guests. Please pick any time you like and we will be happy to pay your expenses—but on one condition. You all stay in our home and Sophia teaches Rose the secrets of her cooking."

Father Passante looks at Sophia.

"Yes, indeed, we will come, and I hope Father Passante can be there as well," Sophia says. "He is like our family and it would be very special for us . . . but you don't want me to teach your wife as well?"

Matt laughs and says not to worry.

"As I said, she is not a friend to the kitchen."

Father Passante raises his glass and toasts.

"So it is agreed; we shall do this trip."

Salvatore says he has always wanted to see the statue of Liberty. Rose suggests that is the perfect place to begin their vacation when they come. Matt speaks of a time when his father took him there when he was a young man and states that he had a good sense of its size as they walked up the stairs inside.

"In fact, the school for the blind that I attended was not far from it, and when we were on the roof our teachers could see it from there."

Father Passante commented that he noticed Matt holding a box when they were seated in the living room earlier.

"Yes, Father, it is something very special. We all went to visit the home of my parents and the current occupant had found it hidden in the staircase. It has a prayer book, a locket with my parents' pictures, and love letters that my parents had written to each other."

"Matt, my sermon tomorrow happens to be on the eternal nature of love. It would be a privilege to speak about this as a part of my message. I would love to see the prayer book, but think it's best to not read the letters. They are too personal."

"Luke, son, would you mind getting the box for Father?"

Luke leaves the table promptly and returns with the box; he hands it to Father Passante and he proceeds to open the smaller box inside and take out the little prayer book. It is a beautiful, hand-bound book of daily prayers.

"I know of this book; my father gave me one as a child. It is very special and the note inside says it was given to your father at the event of his confirmation."

"I would suggest that the contents of this box speak many things about your parents' love of God and each other," Sophia adds, "To have this back with your family is a great treasure."

"Where did your parents live, Matt?" Father Passante asks.

"A man named Major De Francesco lives there now."

"Ah, I know him well. His mind is trapped in a time gone by, but he comes to church every week and always does his best to help others in the village; but his drinking has taken over his life. I am very sure it pleased him to present it to your family."

"He was very proud and saluted us as well," Luke says.

"I would be most privileged if I could hold up

this box tomorrow while I am speaking and bless it at Mass."

Matt thanks him and is most grateful for the offer and says they will bring it in the morning with them to church.

Sophia asks Rose to join her in the kitchen to help with the dessert. Salvatore starts to clear the table; Mark and Luke help the young boy while Father Passante and Matt retreat to the living room, where they are left alone to speak for a while.

"Matt, now that we have some privacy and I know of your situation, is there anything I can do to help you? Certainly this must be a trying time for you and your family."

"Father, I can tell you that I am consumed with rage and hurt. I have a small army of men tracking my wife's steps and an investigation has started to determine who her lover may be. At this point I don't know who he is and I am questioning why I did not sense this coming."

"Matt, have you prayed about this?"

"Yes, but not in the proper way; I was not listening to God but rather shouting at him and I am disappointed with myself about that. All I want is to get her out of my life as fast as possible and learn who this man was."

At this moment Rose was approaching the living room with a tray of coffee but she realizes the men need private time and she returns to the dining room where she tells the others that the men are having a private conversation.

"Matt, it is most important that you speak to your Lord. You must pray about this and put your mind at peace. Our Lord is not a vengeful entity and we must think about what he has taught us. Rage is poison to the soul, and yes, it is a natural emotion, but it is one you must lay to rest as fast as you can or the devil will take over your thoughts."

"Father, I cannot help but think of my first wife and what we had together. I am sure the love letters in my parents' box speak of a relationship much like the one we had with each other and the blessings that love brought. When I got married the second time I heard her voice telling me to be happy."

"Matt, she and your Lord are telling you this now, and you must open your heart and mind through prayer. I will keep you and your family in my daily prayer concerns."

"Having my family and special people around me

is helping to calm me. Father, I promise you that I will reach out to my Lord and will listen carefully as he speaks to me."

The smell of fresh Italian pastries and coffee pervade the room.

"Father, my nose is telling me that we have another calling at this moment and it is now in the other room."

"Yes, indeed, it is; let's join the others, my friend."

Matt takes the arm of Father Passante and they go back into the dining room, where the table has been set and all but Salvatore have a small glass of homemade limoncello at their setting. The Sicilians let their young drink wine, but very often it stops there as the limoncello is too strong.

Mark offers a toast to Sophia and Salvatore for their hospitality. It starts to get late, and Father Passante says he must go to prepare for Mass in the morning.

"Matt, would you entrust me with the box for this evening? I will not open it but will use it in my inspiration and words tomorrow."

"It would be an honor, Father, thank you."

The Messinas offer to stay and clean up, but Sophia thanks them and says that by the time they walk home she and Salvatore will have everything cleaned up and that all must rest for a long day of church and more feasting tomorrow. All bid each other a good night. Once back at the Bianca residence, all retire to their bedrooms. Matt goes to the chair by the window, which is open. It is late at night and all is very quiet with the exception of a distant barking dog. Matt prays for his mind to be at peace and he sits there for a while as he often does. Although he could not see, the windows were like a frame to a picture for him. He stays there for quite some time forming visions in his mind of the world outside before retiring to bed.

Chapter 18:
RUNNING AWAY

Back in Princeton, Dawn and Ken are making their final preparations to leave for Europe. She makes a call to Aunt Pauline so that she can speak to Shamus, as she has no idea if they will ever see each other again or be able to communicate freely. Aunt Pauline answers the call and says how happy she is to have Shamus there. She summons Shamus to the phone, where the two have a brief conversation, as Dawn is beginning to tear up and does not want Shamus to hear her this way. She reminds him to not speak of what has happened and reassures him everything will be all right and tells him to have a great time in Palm Beach.

"I love you, Shamus, you know that, right?"

"Yes, I do, and I love you too, Dawn."

They hang up and Dawn begins to cry; she is nearly inconsolable as Ken holds her in his arms until she is able to collect her senses.

"Ken, I am going to finish gathering up my things; please call a cab for us."

"Sure, honey, I am all set."

He watches her go upstairs and makes sure she is out of listening range. He puts a quick call into his banking connection in Switzerland.

"Thord, this is Ken Wilkenson; I must be brief. I will be in your offices within the next few days; will you be there?"

"I will be here; may I can be of service to you now?"

"No, thank you, Thord; it can wait until I see you. Good-bye."

Ken hangs up the phone quickly and calls the cab. He shouts up the stairs to Dawn.

"Our ride should be here within fifteen to twenty minutes."

"I am almost done up here, do you want to check if there are any other things you may need, Ken?"

"No, I am set, honey; do you need any help?"

"Nope, I will be down in a couple of minutes."

As Dawn walks down the stairs she asks Ken if she should call her lawyer one last time.

"Since it is Sunday, perhaps you could just leave a

message on her machine stating that you have left on a trip and that you will be in touch when it is convenient for you."

"Yes, that sounds perfect, Ken. I will do that now."

"Ok, while you are doing that I will head upstairs and bring the bags down."

Ken opens a small safe in his uncle's closet and pulls out the eight thousand dollars in cash that he had put there for emergencies. Ken had it in the back of his mind all along that he may need this as getaway money to flee the country if he were caught; his worst nightmares were coming true. The cab's horn is heard outside and soon the two are on their way. Ken is looking back at his uncle's old home for what may be the last time as Dawn nervously looks around to see if they are being followed. As they turn the corner on to Nassau Street and pass the university, Ken makes eye contact with the man he had seen in the university chapel the prior day. The man stares right through Ken's soul. Ken is frozen in his seat and quite taken back by the glare in the man's eyes. He is instantly overcome with rushing adrenaline and fear.

"This is such a pretty town," Dawn says.

Ken does not reply; it is as though the man has put him into some sort of trance.

"Ken, hello? Are you in there?" Dawn shakes his knee.

"Oh, ah, yes, sorry, I was just day dreaming. There are so many memories I have of this place."

The two say very little on the ride to the airport as they don't want to speak in front of the driver and they are both nervous about their situation. As they get close to the airport Ken pulls his passport out of his wallet and laughs at the sight of his picture.

"I cannot believe I ever let my hair grow that long, not to mention that cheesy mustache. I could never have that looked that way in the office today."

"Yes, I am sure those tight asses would not approve. In fact, I don't think I recall anyone having facial hair in any of my husband's companies," Dawn replies.

The cab driver over hears that remark but tries not to let them know he heard it. As the cab pulls up to the curb Ken hesitates getting out.

"You two relax for a second. I will get your things out of the back but I will have to leave quickly; the airport police keep this drop off zone clear."

Ken is noticeably nervous and assures Dawn that it is just last minute jitters. Soon, they are through the security checks and at their gate waiting to board. The

New York City skyline can be seen in the distance. Ken gets up from his seat and walks to the window and places his hands on the glass, almost as though he were touching the city.

Dawn cannot help but think how Matt often did something similar with windows. She gets up and puts her arm around him and takes in what may be their last look at the United States. Their plane is called and the two board holding hands. A flight attendant asks if they would like anything before they take off and both order vodka martinis. When she brings them their drinks, the attendant comments on what a beautiful couple they make. Dawn thanks her and Ken simply returns a smile. As the plane is taking off Dawn realizes that they have not made hotel reservations.

"Ken, did you get us a place to stay? In all our rushing I don't recall doing that."

"No, I didn't, but the bank has preferred relationships close by, so we will be taken care of. I suppose it's one of the perks you get when dealing with private banking in that country."

Ken and Dawn are both thinking to themselves that they have forgotten other things that may expose them. The drinks keep flowing and soon the two fall asleep holding hands. The flight attendant puts a blanket over the sleeping couple. About an hour prior to landing a routine announcement is made and they are awoken. Dawn has never been to this country and gazes out the window as they descend.

"This is a beautiful place, Ken. I always wanted to come here." T

"The people are just as beautiful as the place they live in, Dawn. So much has happened in their part of the world over the last one hundred years and yet this country and its people have been able to remain virtually unaffected by war. It is as though the country were a remote island."

"An island sounds nice, too, Ken."

Ken just smiles and begins to think he has lost his mind and that Dawn has lost any common sense. In fact, he is thinking she is simply a nut case and it has taken this tragedy for him to realize it. Once the plane is on the ground and they have gotten through customs Ken asks Dawn to wait a minute while he makes a call to his banker for hotel reservations.

"Thord, this is Ken; I am sorry to bother you but I did not make a hotel reservation."

"Oh, not to worry, my friend. Do you recall the chateau you stayed in the first time you were here?"

"Yes, it is an amazing place."

"Ken, I can have a driver there for you in less than thirty minutes if you can wait. It is too far for a cab anyway."

"Great, Thord, I really appreciate that."

"His name is Leif and he will be at the baggage area with a recognition plaque with your name."

"Great, our flight was Liberty Air number 369, arrival from JFK." Ken turns to Dawn. "We are all set. You will be in heaven when you see this place, and the banker is sending a driver."

Since they have to wait a bit for the driver they have a few drinks at a bar inside the terminal. A short time later they are greeted by a smartly attired driver and they are loaded into a very well-appointed Rolls Royce. The views up to the chateau are breathtaking. Snow-capped mountains, bucolic villages, and winding roads make for one of the most beautiful vistas on the planet. Dawn comments that she has never been here and cannot believe how breathtaking it is. The driver comments that he was born and raised there and that the views still impress him everyday.

"The mountains look different every time you see them. Not only do they change dramatically from season to season, but also from hour to hour as the sunlight and clouds seem to change their colors and shapes," he says.

Ken looks away from Dawn and as he looks out his window he says, "It is interesting how things can seem much different than they are."

The driver dismisses the remark but Dawn is clearly not happy with Ken's cynical tone and comment. There are crystal bottles full of the best liquor in the back of the car. Dawn offers to make Ken a drink but he declines, citing that he will need to be focused on his bank meeting.

"Ok, suit yourself," and she pours herself a double vodka on the rocks, which she finishes rather quickly. At this point no one is speaking in the car. The tension is obvious and Dawn pours herself a second double of vodka. As they come around a curve high in the mountain the chateau comes into view. It is like something out of a storybook. The home was built in the early 1900s by a European industrialist for his large immediate and extended family. After the Second World War the bank bought it from the family's estate for client use and meetings.

As the car approaches the gates are open and friendly chateau staff employees are at the door ready to greet them. One of the butlers advises them that Mr. Thord Van Clief will join them in the conservatory once they are settled and freshened up.

Once they are shown to their room Ken suggests that Dawn stay behind for the men to discuss the bank account.

"Dawn, I don't want you in the room; you have been drinking a lot and I don't want things to slip out."

"You sound just like Matt, always trying to control anything around finance."

Dawn's voice becomes stressed.

"We are in this together, Ken, and I am coming to this meeting."

Ken draws close to Dawn and takes hold of both her wrists.

"Yes, we are in this together, and you will do your part by not showing up to a meeting drunk and slurring your words. These people are serious and they do not have tolerance for nonsense and if there is any hint of wrongdoing or lack of decorum we can jeopardize years of planning. Look how one little slip from your brother brought us up to this point. Now you sit down and stay here. I will be back shortly and we will get this all on track."

Dawn sits down in a chair and begins to cry.

"We are going to go to prison; I just know it," says Dawn in a tearful rage.

"You are proving my point, Dawn. You are too emotional for this meeting; just stay put and I will be back soon."

Ken leaves the room and makes his way down to the conservatory, where Thord is waiting with an assistant. Thord is a stately man of about fifty-five years with salt and pepper hair.

"Well, my good man, it is good to see you, Ken. I understand you arrived with a beautiful lady."

"Yes, I am traveling with a friend this time. We would like to spend some time in Europe. In fact, I am thinking of settling here and have identified a few homes that I would like to consider purchasing."

Ken has come up with a neat story to withdraw some funds. He really doesn't need to have a reason, but would rather keep things moving without hesitation or cause for concern.

"Thord, it seems that one of the owners is a bit a eccentric and wants half of the money in cash up front. Please have five million brought to me here this evening and I will leave the rest on deposit with your bank."

"Most certainly, Ken, may I have you and your beautiful lady to dinner this evening with me and my assistant?"

"That is very kind of you, Thord, but we shall be taking dinner in our room this evening as my lady friend is not feeling well and is a bit exhausted from our journey."

"That is most understandable, Ken. I will have my assistant bring the money to you."

"Thank you, Thord, but I just recalled I have some things in my safety deposit box that I would like to pick up so I will come by in the morning to get the money and my personal effects."

Ken wants to get the money without Dawn because he is planning to make a run for it without her. The men shake hands and Ken returns to the room where Dawn has fallen asleep on the bed. She is awakened by Ken's return to the room and is clearly in a daze from her drinking and travel exhaustion. Ken tells her to stay relaxed and he will have dinner brought up to the room.

"Thank you, Ken, I am just too tired to go anywhere, and besides, this is such a beautiful room and the views are wonderful."

"I have had the salmon here. It is the best I have ever had; why don't we get two of those?"

"Sure, Ken, that will be great. So tell me, how did your meeting go?"

"Well, it seemed that Thord was a bit curious about my 'lady friend,' as he put it, but it went well. I just tried to keep the conversation as light as possible to remain on task with the business at hand. I made an appointment for a withdrawal early tomorrow morning so we will be all set then."

"Ken, I am so happy we are here together. For some reason I feel like nothing can touch us or do harm to us here. Perhaps it's the secluded mountain location."

"Yes, I feel the same way my love; all will be just fine. The drama of the last few days has been too much for anyone to bear, so we just need to relax, Dawn. I know you have had some drinks, but I think this calls for a nice bottle of champagne, don't you?"

"Sure, please call in our order. I will take a shower and freshen up."

While Dawn gets into the shower Ken looks through his bag and locates some sleeping pills. He quickly takes a few of the tablets and places them inside his pocket and waits for the right time to lace her drink with the powder from inside them. The food is delivered to the room while Dawn is still in the shower but Ken doesn't put the powder in her drink. His plan is to wait for breakfast so that the drug will have her out for most of the day and by then he will be long gone with the cash. He wants nothing to do with her and feels as though it's best to just get a fresh start where no one knows him and he can disconnect as much as possible from any ties—especially ties to the Messinas. He is not so stupid as to think the Messinas will not be after him, but rather his plan is to stay as far away as possible, especially with Dawn's drinking, which is a clear exposure to risk.

Dawn comes out of the bathroom with nothing but a towel on, sashays by Ken, and faces out the window, were she drops the towel away from her naked body while turning her head to him, exposing her bottom. Ken thinks to himself, "What kind of mad man would leave this?"

She knows Ken has been upset with her all day and clearly does not know of his dastardly plan. Ken enjoys the show she is putting on while he pours the two of them champagne.

"Well, it looks like this food may have to wait, Dawn, doesn't it?"

"Oh no, Ken, we will eat now. I am the dessert."

The naked Dawn eats every bite of her dinner seductively and slowly. Ken continues to refill the champagne and thinks to himself that he may as well make this last night a good one. They look at each other with hollow eyes. Ken feels as though he was drawn into this mess by an evil seductress, and she feels as though he was her way to extract more money from her unsuspecting husband. They know now that, really, it was not each other they loved but rather the excitement and the money. The attraction was merely sexual and now it has become an emotional tug of war waiting to explode. They could see right through each other's soul at that moment and the recognition both intrigued and disgusted them. The emotion of intrigue won over as more champagne flowed. Ken says he has had enough dinner but saved lots of room for dessert.

"You have to wait a moment; it has to cook a bit. Stay there," she tells him and she walks slowly over to the bed. She lays her naked body down so as to give him a full view of her entire front, then with erotic precision rolls over to her stomach and lifts her backside up and down several times. Ken cannot take it any longer and runs to the bed, where after several hours the passion ends in the two falling into a deep sleep.

As the sun comes and blankets the room with a warm glow, Ken is the first up and walks into the other room where he calls for in-room breakfast. He instructs the service to not knock on the door but to enter quietly as he will be waiting and doesn't want to wake Dawn. He gets dressed and waits by the door, where the food arrives rather quickly. He asks the service person to set the table next to the window so a view can be enjoyed for the meal. He tips the man generously and closes the door behind him. Ken makes sure Dawn is still sleeping and goes back to empty the powder from several of the sleeping pills into Dawn's juice. This delivers her triple the dose that he knows she typically would take to make her sleep for the night. Ken then goes back into the room and kisses Dawn on the back of the neck a few times. He tells her that he has arranged a wonderful breakfast for them.

"You are a helpless romantic, Ken."

"Yes, I know I am, and I cannot stop being that way."

"I will join you in a minute; let me freshen up and I will be right out to join you, Ken."

He thinks he should add more of the drug to the juice but is afraid of being caught so he just waits by the window for her. Dawn returns with a bathrobe on.

Embroidered on it is the crest of the chateau. It has a regal yet sinister quality to it. In fact, it has remnants of a swastika-like cross at its center. She notices Ken staring at it.

"What is wrong, Ken?"

"I cannot help but think that the crest almost looks like the symbol of German tyranny."

She looks down at it.

"You are being silly, Ken; it is just a cross."

"Well, if you say so, but that is what my eyes see and I never really noticed that until now. I have been here several times but never thought of it that way until I saw it on you."

"What the hell are you saying, Ken?"

"No, honey, I didn't mean it that way; I just never noticed it. Let's eat our breakfast before it gets cold. I know how much you love fresh-squeezed orange juice, and they actually did it right here for us as they set up the table."

She takes a taste and says it is the best she ever had and suggests that it would make a wonderful mimosa. She finishes half the glass and asks Ken to open a bottle of champagne to add it to the remaining orange juice in her glass. He cannot believe his luck; the combination of the two things will really knock her out. She finishes that first mimosa in very quick order, as Ken does with his, to keep the momentum going. At this point they haven't had a bite to eat and the drug is well on the way to doing its job. Ken makes two more mimosas as Dawn picks up a croissant and begins to butter it.

"I am feeling a bit dizzy. I think we had too much to drink last night. Maybe we should just eat and take a break from the champagne for a bit."

Dawn's head starts to bob up and down as she tries to fight off the effect of the drug that is making her drowsy.

"You know, I think I need to rest a bit more."

"Sure, love."

Ken helps her back to the bed where she is fast asleep in a few moments. Ken sits in a chair beside the bed to make sure she is truly asleep and that the drug has taken complete hold of her. He thinks that he must move fast, yet very methodically. He can't be seen down in the lobby or call a car right away, as that will look strange to the staff. But then he thinks, "Oh well, who cares?"

His mind is racing and he tells himself to calm down. He walks over to the table and picks up the croissant that Dawn had just buttered and starts to eat it on his way to the bathroom to take a shower. He turns on the water to let it run a bit to get it hot and walks back to the table to pour a cup of coffee, all the time still eating the fresh pastry. He pauses for a moment and goes to the writing desk, where he leaves Dawn this note: "I am sorry, but I must go on without you. I love you but will never be in love with you. I cannot forgive myself for what I did to Matt and his family nor can I continue to look at you. I left a key for the safety deposit box in your purse. I will give Thord and his people instructions to let you into it. You are not to discuss any of our business while there. I will leave half of the money in cash in it. I will not be coming back to the United States and you will have what you wanted all along: your freedom and more money. Good-bye, Ken."

He returns to the bathroom and on the way leaves the note on the pillow next to Dawn. He takes a moment to look at her and tears well up in his eyes. He then goes to his briefcase and retrieves the second key to the safety deposit box and places it in her purse. He looks at the clock and notices that he needs to be at Thord's office in an hour so he expedites his shower and gathers his things. Before he leaves the room he checks on Dawn one last time. He rereads the note one last time, and then as he opens the door to leave he notices a dozen pink roses in a vase near the couch. He could not help but remember how Matt came into the room with roses like that when they were caught. He takes them out of the vase and drops them on the ground, just as Matt had done that fateful day in New York City.

He doesn't call for a bellman as he wants to leave without interruption. A car sent by Thord's office was already waiting in the driveway and is called to the door. He informs the concierge that his lady friend is not feeling well and that she is exhausted from her trip and to please advise the staff not to wake her or enter the room until she requests service herself.

"Yes, most certainly, sir; enjoy your day. Is there anything else we may do for you?"

"No, thank you, I will be back later this afternoon."

As he is getting into the car and the driver is loading his bags he looks back at the concierge and they both realize he is loading his things and doesn't really give the appearance of returning. Ken thinks to himself that he is already making tactical mistakes. The driver is the same gentleman that dropped them off the day before.

"Sir, will I be bringing you back to the hotel later or

to another destination after your meeting at the bank?"

He thinks quickly and says he may have to make a day trip but will know in an hour or so and that is the reason he has his bags, since there may not be time to get back to the hotel.

"Certainly, sir, I will wait outside the bank and await further instruction."

Ken is greeted in the bank lobby by Thord's assistant. This is no ordinary bank; it looks like the museum of fine art in Manhattan. Tapestry and fine oil paintings cover the walls in the greeting area where the ceilings must be at least thirty feet high. Ken meets with Thord in his office, where the cash is waiting for him.

"I am glad we will not be losing you as a client, Ken. You mentioned that you will be leaving half your funds in our safe keeping."

"Yes, that is correct, Thord."

"I wish you well with your home negotiations."

"Thank you, Thord; in fact, I will be taking a day trip there today and while I am away my lady friend would like to have access to my safety deposit box. I have given her my second key."

"Of course, Ken."

Thord motions to his assistant to get the necessary document for Ken to sign, which allows Dawn's access.

"I am sorry, Thord, I must be efficient with my time as I will be meeting with the owner of the home and have to do a bit of traveling to get there."

The assistant returns with the form and Thord hands Matt the briefcase loaded with cash. Ken asks Thord for access to the safety deposit box before he leaves. Thord wishes him well and the assistant leads Ken to the private area of the safety deposit box.

He is left alone, where he leaves half of the money from the briefcase for Dawn and takes some personal effects that he had been keeping there, including some gold bars he had purchased over time. He summons the assistant and Ken is led back to the lobby and exits the bank. The driver is patiently waiting and opens the door for him.

"Sir, where may I take you now?"

"To the train station, please."

They pass by the chateau on the way there and as they drive by Ken notices the crest at the gate and cannot help but see the evil in the symbolism. He wonders how he truly never noticed it before and stares at it until it is out of sight.

Chapter 19:
CHURCH AND FAMILY

Morning is breaking in the home of the Biancas, where all are preparing for church. This morning's breakfast will be a short one, as Mass begins at 9 a.m. and the feast follows directly after. Rose comes to her father's room as asks if she can be of any help to get her dad ready. He instructs her to call the hanger to remind the flight crew they are invited for today's party and to come ready to eat and enjoy the day.

"Please tell them not to worry about drinking, as their will be no need to fly for a while."

He has made the decision to stay and enjoy some travel in the area.

"Rose, please come back in a few minutes; I will be dressed and ready to go shortly."

As Matt puts on his suit he feels something in his inside jacket pocket. It is his St. Joseph's medal that he has carried since he was a small boy. His mother had given it to him for blessings while traveling. While he was in the school for the blind he would keep it under his pillow at night. He is surprised, as he thought he had left it home, and realized that Damiano had likely placed it there for him as he has done in the past for other trips. He sits in the chair by the window and holds it close to his heart with both hands. In a prayerful whisper he says, "Mom and dad, I know you are here with me right now. I pray that you are in God's hands and will be with me in church today. My greatest blessing is your wisdom and my family and to have them here together in this place along with your box is confirmation that our God is watching over us all."

Rose comes to Matt's room and she brings him to join the others. The Biancas and Messinas gather for coffee and pastries and head off to church, which is a short walk away. As they leave the home, many others in the village are doing so as well and the streets fill with people in their Sunday best. The church bells begin to ring as they turn the corner to the street the church is on. Matt tells his children to take note of the plaque on the first pillar on the right side once inside the church. As they enter the church they all place their fingers in holy water, genuflect, and make the sign of the cross. Luke locates the plaque and Rose translates it: "This plaque is dedicated to the Messina Family, who gave so much of themselves to build this church. Without their faithful dedication to God and this village the construction of this house of God would not have been possible."

Matt reaches for the plaque and places his right hand on it and places his left hand on his heart. The touching moment is noticed by many of the parishioners. An usher takes the family to the front pew, which has been reserved for the family. This church is an amazing structure; it was built in the traditional style seen in many seventeenth and eighteenth century houses of worship. Every inch of the inside walls and ceilings have been beautifully detailed with frescos and gold-leaf trim. The church survived many challenges without any damage whatsoever, including World War II, which in and of itself seemed to be a confirmation of this being a blessed place.

The organ begins to play and the procession of the altar boys, assistant pastor, and Father Passante makes its way to the altar. Father Passante is carrying a Bible and the Messinas' box. It is now a part of the mass, reserved for the sermon.

Father Passante walks to the center of the altar and addresses the parishioners from the center aisle, adjacent to where the Messinas are seated.

"My dear followers in Christ. Today we are blessed to have the Messina family here with us today. As many of you know, this place of worship and the treasures of its space would not have been possible without their ancestors. Today, I have a message of love to share with you. Love has no end and no limits to what it can do. The Messinas' love of God is the energy and power that built this church and it is your love of God and each other that keep this place vibrant and lasting for us and all generations to come. You may have noticed that I am holding a small box. Recently, one of our fellow parishioners found this in the former home of the Messinas when he was doing renovations. The box contains a prayer book and letters of love. These belonged to the parents of Matt Messina, who is here with us today. My friends, the message here is that Matt's parents treasured the word of God, the love of God, each other, and their family more than anything. For us to have found this on the eve of their arrival to our village is no accident. God wants us all to see the message in this. We must recognize the significance of this and that we are here to carry on and preserve the word of the Lord. We must remember his word each and every minute of every day. I will now return the box to its rightful owner."

Father Passante passes it to Matt. At that moment Matt stands.

"Father, if I may: I would like to present this to the church where I think it should be sealed forever and kept in this hallowed place. I am sure with prayer you and the parishioners will find a suitable place where this can rest in this space. I feel my parents' presence here today and feel this is something they would be pleased with."

"Matt, it would be an honor; thank you."

All in the church are touched by this tender moment. Major De Francesco stands and salutes the Messina; all in the sanctuary know is he a bit off and they smile. At the conclusion of the service Father Passante announces that coffee hour will not be held today; instead, all are invited to the square, where the Bianca family will be hosting a Sunday feast for all that would like to attend. The festivities will begin at noon and there will be music in addition to wonderful food. The ushers come to the front of the church and lead the Messina family out directly behind Father Passante and his processional.

Back at the square Alfredo has been setting up for hours with the help of his staff. Giovanni was also there with the help of friends that he had enlisted from the area. He tells his friends that Rose is off-limits.

"She is my cousin and my uncle is a very powerful

man. None of you clowns are suited for her, but I know when you see her you will try. I am telling you now, that is not going to happen."

One of the men replies, "Yes, sir, we will be sure to forget we are Italian, too, if you want," and all laugh it off. A band is set up next to the fountain and begins to test their equipment. They have come with a locally famous singer who has a voice that is reminiscent of Frank Sinatra with looks like Dean Martin. He is quite the performer and will lead the day.

Soon people are starting to fill the square. In the distance the flight crew from the Messinas' jet makes their way into the village from the hill above. Luke spots them walking in and says he noticed them coming down the hill.

"You guys didn't walk all the way here, did you?"

"No, we had a cab drop us off at Alfredo's place first, since your father spoke so highly of it. We thought we would check it out, but it was closed."

"Well, that is because they are all here setting up for us."

"Well, Luke, your family never ceases to impress us," Blake says.

They all thank Luke and head over to meet their hosts and greet the rest fof the Messinas.

The band starts to play and the party gets into full swing quickly. There are hundreds of people there, and although not asked to do so, most brought even more food and wine. There seems to be more food than at the best tailgate party back in the United States. Giovanni cannot help himself and the old show-off in him comes out and soon a crowd has gathered around to watch his dancing skills with a young lady he has summoned from the crowd. A circle has formed around the dancing couple and when they are finished all join hands and do the Passu Torrau and the Tarantella. Sophia has been watching Matt the entire time and her son notices this. He suggests that Sophia ask Matt for a dance.

"Son, that is not appropriate, for a lady to ask a man to dance."

"Well, then why not ask Matt if you can bring him something to eat."

"Yes, that is a fine idea, my son; I shall do that."

Sophia walks over to Matt's table, where his family had been taking a break from the dancing, and she asks Matt and his family if they would like to have some special treats that one of her friends made.

"We would all love to try them," Matt says. "If they are half as good as your cooking it would be very special."

"Wonderful, Salvatore and I will be right back."

Women have a way of being able to read signals and Rose can clearly understand that Sophia has taken a fancy to her father. She does not say anything but keeps a watchful eye on this. Sophia and Salvatore return with several small plates of delicacies.

"We have brought you scallops with black beans, grilled calamari sautéed with lemon and virgin olive oil, and some local Bronzino with a wonderful secret preparation of local herbs."

Matt stands and asks them to please sit down and enjoy the food with them. Mark comments that these are some of the best things he has ever tasted. The food and wine are flowing and the music slows to a slow pace and comes to a stop; the bandleader asks those with a partner to join for a slow dance. Matt asks Rose if she would like to dance with him.

"Of course, father, I would love to. Several couples are soon dancing and halfway through the first song Rose suggests that Matt dance with Sophia.

"Dad, would you mind? She has been so good to us and I can see her watching all the couples and this must be hard for her."

"Yes, that is a very kind thought."

Rose motions to Sophia and at first she doesn't get up.

"I think they want you to join my father, Sophia," Luke says.

"Well, it is has been a long time for me; I don't know. I have forgotten how to dance."

"Mom, please dance; that would be nice," Salvatore says.

So with that Sophia walks over and Rose takes her hand and places it in Matt's hand. It seemed as though time has stopped in the square as all take notice of this. One couple at a time drifts off the dancing area and before long it is only Sophia and Matt. Another song has started and they are the only two remaining. The entire crowd is watching this moment.

"I have the impression that everyone else has left the dance floor, as I cannot sense another person around us," Matt says.

"Yes, Matt, it is just us."

"I hope you are fine with this, Sophia."

"Yes, Matt, very much so, and thank you for asking."

The sapphire blue Sicilian sky, the whispering

sounds of the fountain, the perfect melody being played, and the flawless dance made this the most romantic moment anyone could ever hope for. As the music slows to a stop Matt reaches for Rose's hand and kisses it gently. The crowd breaks into applause and the music picks up the pace as many come back to dance.

Sophia leads Matt back to the table, where the family has been watching. As they approach the table everyone is getting back up to rejoin the dancing.

"Dad, you are not going to sit down now, are you?" Rose says.

It seems as though hundreds are now dancing to popular Italian dance songs. The flight crew, Alfredo and his staff, the villagers, and even Father Passante are in full swing. The party continues well into the evening and it seems like there is an endless supply of food and wine. As midnight approaches the band asks for any last calls.

"Do you guys know any Andrews Sisters?" Matt shouts out.

The bandleader says he is not sure.

"You know, like the "Boogie Woogie Bugle Boy," Matt shouts back.

"Ah, yes, of course we know that one."

Matt grabs Rose's hand.

"Do you remember when you, me, and mom would do this dance in the kitchen when you were young?"

"Dad, of course I do."

"Well, are you up for it?"

"Of course, let's do this!"

Most of the guests are still there and gather around to watch Matt and Rose. The beginning of the song starts with a bugle call from the band's trumpeter, and with that, Major De Francesco , in full uniform of course, stands up and salutes. This is a bit of a strange and awkwardly funny scene, as when this song was written the Americans were fighting the Italians and Germans in World War II. Matt's mind races back to the first time he met his dear wife Rose and heard her trio sing this song back on the Jersey Shore Boardwalk.

Sophia, Salvatore and Matt's boys are standing together watching this play out. The crowd starts to clap in unison to get the dancers in synchronous motion. People are whistling and shouting as they break out into dance. They had done this many times before and it came back to them quickly. Matt felt his wife's spirit there and his motions seemed as automatic as Rose's. Sophia comments on what a great man and dancer

Matt is. Mark, sensing her emotion for his father and not particularly favoring it, says in a curt tone, "He certainly is; my mom used to sing and dance to that song when she was in a talent act during the great World War II. It was my parent's favorite song."

Luke whispers in Mark's ear, "Leave it alone, brother. Dad is having a great time and this is just a party."

"Luke, can't you see what is going on here? She is a poor woman after Dad's money."

"Mark, you are always suspicious of everyone that comes into Dad's life."

"Well, case in point, was I not that way about Dawn?"

"Well, yes, but please just drop it and let's not make a scene here. This party is going so well and I have never seen dad happier since mom died. Mark, I think you are wrong about this woman; let's not worry about it. We have bigger fish to fry with Dawn and helping dad get back on his feet."

"Well it looks like the feet are bouncing back pretty well right now."

The last few notes of the song are played and Father Passante takes the stage to thank everyone for coming and bringing such wonderful things to the event.

"God Bless you all."

Chapter 20: TEMPTATION

Back in the Alps, Ken's train is making its way through the snow-covered mountains; he notices none of it. He sits with a blank stare and thinks to himself that he is a fugitive in a foreign country and considers what the implications of that are and the likely impact they will have on his life. He vows that he will never have another drink of alcohol or look at another pretty woman as long as he lives. He thinks to himself that this is not how he was raised, and wonders how his decisions could have been so misguided.

As he is having these thoughts, his striking good looks catch the eye of a beautiful blonde woman, and as he turns and looks around the train they make eye contact. He turns away to look out the window. She has a classic beauty and is very well dressed. She has just finished a martini and another is being served. Ken keeps trying to ignore her but there is something beyond her looks that keeps his curiosity piqued. He gets up to walk to the water closet, where the coach master pulls a curtain divider closed at his service area and asks Ken not to stare at the lady. Ken says he was doing nothing of the kind.

"Well, sir, she is a very famous actress and I have been asked to keep her space private."

"Well, you don't have to worry about me. I am in no mood to look at her or any other woman for that matter. If she is so famous and wants her privacy why is she not in a private car?"

The curtain swings open. She had gotten up as well, and heard the entire conversation.

"It is all right, gentlemen. This man was of no bother to me. In fact, it was I who was staring at him. Hello, my name is Alessia Eversoll; pleased to meet you."

"Nice to meet you; I am Ken."

"Would you like to join me? The views are so beautiful, and it's a shame not to share this ride with someone."

"Well, you make a good point, Alessia; I would love to, that is very kind of you."

After she is reseated, Ken joins her at her table and he is face to face with the most beautiful woman he has ever seen. This is an interesting conundrum after having just sworn off women and booze.

"Ken, I often ride this train and never stop marveling at the views; I don't think I have ever seen you."

"I have never been on this route, and yes, it is truly beautiful."

A server stops and asks if he can offer something to the couple.

"I am fine."

"Perhaps the gentleman?" the server asks.

"No thanks, I am fine."

"Perhaps a glass of ice water, sir?"

"Yes, that will be fine, thanks."

"Not a drinker, Ken?"

"Well, I need to be focused on some things."

"Oh, I see; have I disturbed you?"

"No, Alessia, in fact this is nice and I appreciate your company. I have been on a bit of a whirlwind business trip and have not caught up with the time changes from the United States."

"Oh, I understand completely. I travel there often, mostly to New York and Los Angeles."

"So, tell me, Alessia, it must be fascinating to be an actress."

"My favorite work is stage but of late I have been filming movies."

"Please forgive me, Alessia, but I am not familiar with your work. I have not been to a play in a very long time nor do I see many movies."

"Actually, Ken, it is refreshing to meet someone who doesn't recognize me."

"I didn't mean to insult you in any way."

"No, in fact, you did just the opposite. Most people feel a bit awkward when they meet me and that is uncomfortable. May I ask what you do, Ken?"

"Well, I am actually between things right now and thinking of working for some private investors. I have been in the financial services industry for many years."

"Well, this is an interesting country for that sort of work, Ken."

"Yes, you might say that."

"What is your destination today, Ken?"

"I am going to Lucerne and not quite sure how long I will be staying."

"How perfect, Ken, I am hosting a dinner party tomorrow evening for a small group of friends and I hope you don't think this is too forward, but would you like to come?"

"Would you mind if I got back to you later today or in the morning? I may have a business commitment."

The coach master overhears this and can hardly believe his ears. Every man in Switzerland would jump at the chance.

"Most certainly; it would be nice to have a fresh face at the table and my friends are fascinating people. I am sure you will enjoy the evening if you are able to come."

Ken's mind is rushing; how could he be so lucky and yet not be able to follow up on her offer? He knows Dawn will be over the top with anger when she wakes up, but he is thinking she will just take her cut of the money and have her own pity party, as she must go into hiding herself. He knows that if he were to be seen with such a high profile woman that it may get out into the society papers. This is a terrible problem for a man trying to disappear.

"Here is my card, Ken; do you happen to have one?"

"I am sorry, when I left my last firm I did not bother to have a personal card printed."

"Where are you staying? I can have a driver come to get you."

Ken thinks quickly.

"I may stay with my clients, but will decide that later today. May I call you when I am certain of their plans and where I will be located?"

"If you like you may bring them as well."

"Thank you, but they are not the social type. I am not sure that I will be able to break free; I am meeting them today and all day tomorrow, but frankly, I will likely need a break from all that."

The coach master announces that they will be arriving at the Lucerne station in ten minutes.

"Ken, there is something about you that I find very interesting and I don't make it a practice to invite people that I have just met to my home."

"Alessia, I am very grateful for the invitation and hope that I can extract myself. If this doesn't work out I hope that we may meet another time."

"Most certainly, would you mind giving me your last name, since you did not have a card? Here is another one of my cards; just write it on the back."

Ken writes his name, all the time knowing he is

now laying a trail for the authorities; however, he remains confident that Dawn will just fade away and not pursue him. It is the Messinas that he worries about more.

"My driver will be waiting for me at the station; would you like us to drop you off at your meeting?"

"You are too kind, but I would not want them to see me arriving in that fashion. They are very quirky people."

"I understand, Ken; the type of clients you have must take great care in their privacy. In fact, my security people and those trusted with my care would be losing their minds right now with the invite that I have extended to you, but I don't care. I think I God has given me a gift of judging people."

As the train slows to a stop the coach attendants start to gather Alessia's several pieces of luggage.

"This was a wonderful ride, and if I might say, it was almost surreal," Ken says.

The two exchange a hug and she wishes him well at his meetings and hopes that he contacts her soon. She is first to exit the train and turns to waive as she prepares to enter her waiting limousine. Ken stares at her as she walks away with a sexy gait. Her high heels, tight dress, and perfect shape are too much for any man to resist. The coach master has seen and heard much of their conversation. He wishes Ken well and good luck with his endeavors with Alessia.

"Well, sir, that is kind of you, but frankly, I have found that beauty and money can be confusing things, especially when mixed in the same glass, if you catch my drift."

"Well, I am not sure, sir, but either way you are a lucky man and I wish you the best. Will you need some assistance with your bags today?"

"No thanks, I will be all right."

Ken tips him well as he disembarks the train and makes his way to the station after making sure Alessia's car is well out of sight. The stress of the last few days has taken a noticeable toll on Ken. He looks across the street from the train station and notices a café, which appears to be a quiet place for him to get a cup of coffee and take a break for a short time to gather his thoughts and plan his next moves. He notices a telephone booth outside and decides to call the bank. He informs Thord to expect his lady friend sometime soon and warns him that she has a curious mind.

"She is only to get access to my safety deposit box, and I know this should go without saying, Thord, but she is not to know anything of any account I may have with you."

"Yes, of course, Ken; is there anything else I need to know?"

"No, not at this point. I will be very busy with my personal business. I will remain in contact with you and keep the remaining assets in liquid instruments should I need to get access to some of it."

"I understand, Ken, and good luck with your business."

Ken thanks Thord for his professionalism and returns to his table in the café to ponder his strategy.

Dawn is starting to wake up from her drug-induced sleep and reaches her arm out to the empty side of the bed. Her eyes are barely open and she calls out for Ken several times.

"Honey, where are you? I must have had far too much to drink last night; I can barely get my head off this pillow."

Dawn's drug-induced sleep so was deep that she had forgotten she had already been up for breakfast, so she is confused by the remnants of the breakfast still in the room. She manages to get up and go the bathroom, but on the way notices a note on the writing desk. She doesn't pick it up at first but decides to the bathroom. She washes her face and looks at herself in the mirror and cannot believe how bad she looks and feels. On her way back to the bed she looks back at the note and decides to sit down and read it.

"That son of a bitch! That freaking son of a bitch!"

She rips the paper in half and goes into a rage. She pounds the table and breaks a lamp and some decorative statues that adorn the desk. She hurts her hand by pounding it on the desk and goes back to the bed and cries herself back to sleep for a brief time, then awakens to the reality of her situation. She is alone in a foreign country with no friends or family and is now a lonely fugitive. Dawn's evil core grips her and she goes into what Ken used to call "Dawn Mode." Once she is determined to do something, there is nothing that can stop her. There are no more tears, just an evil bitch that is now on a mission. She decides to rally by getting herself dressed to impress. She selects her favorite black dress and shoes and whispers to herself in an evil tone: "Even a blind man loved these."

As she is putting the finishing touches of her make-up on she looks into the mirror and says out loud, "Who do these men think they are screwing with? I don't need anyone; I am going to take care of business my way. I don't need any man or anyone."

She stares at herself in the mirror.

"No one loves you more than me, honey; I will be just fine. There are plenty of Ken's and Matt's out there to fuel my needs. I can chew them up and spit them out."

Dawn calls the bellman and requests a car be brought for her to go to the bank for a meeting. She takes her best Hermes bag and briefcase and heads down to the car. Everyone in the lobby takes notice of her intense beauty and chic but dastardly countenance shadowed by her dark sunglasses.

Once at the bank, she is greeted by Thord's assistant, who tells her she was expected and will be escorted to Ken's safety deposit box. She tells the assistant that she would like to meet with Thord but is told he is in a meeting. Dawn insists that he take a break, albeit for only a moment, to speak with her. Thord excuses himself and comes out to speak with Dawn.

"I am sorry, I must be short, madam, as I am attending a client meeting."

Dawn asks that the assistant leave the room. Thord nods and the assistant leaves. Dawn, still with her sunglasses on, as she is worried that her eyes look tired, says in a very stern but quiet and controlled tone, "Is there anything I should know about, Thord?"

"I don't understand your question."

She fumbles for a cigarette in her bag and Thord lights it for her. She takes a drag from it and blows the smoke right in his face.

"Let me just say this, Mr. Banker; don't screw with me."

"I am sorry; I must to get back to my meeting, madam. Is there anything of substance you would like to speak about? If not, I don't have time or need for any discussion such as this. I run a very professional organization here."

"No, I am done with you for now, Thord but you have not heard the last of me; now go back to your little meeting and have your assistant take me to the safety deposit box."

"Yes, of course, madam, wait here and you will be taken care of."

The assistant returns and takes her to the box, where she is offered a private room. Dawn opens the box to find five million dollars in cash in large denominations. She fills her briefcase and bag and leaves to go back to the hotel. Upon her return to the hotel she is greeted by the concierge, who asks her if there is anything she can do.

"Yes, in fact, please have your best caviar and vodka brought to my room. I will be taking lunch on the veranda."

"Yes, of course; would you like it right away?"

"Yes, please, that would be nice."

As she walks to the elevator she says to herself, "I am going to rack up the bill for my friend Thord as a thank you."

By this time the chateau staff receive a call from Thord asking if she has returned and to deliver a message to her if so. She is welcome to stay another evening as the bank's client as a courtesy, but after that all expenses are her own should she decide to stay. The lunch is delivered to the room with the note from Thord. As she sips the vodka and enjoys the caviar she notices the folded note on the tray. She stares at it for a while while she continues to enjoy the view and the delicacies. She takes a pause and lights a cigarette then walks out onto the balcony with the note in her hand. She opens and reads it, rolls it into a ball, and throws it over the stone wall and into the trees far below. She laughs.

"Men and their silly notes and idle threats. I am going to make Thord pay, one way or another."

She calls down to the concierge and tells her how wonderful the lunch was.

"In fact, it was so good, please bring up another one exactly the same."

She knows that lunch is at least one thousand dollars or more and is going to ride this gravy train as much as possible.

Back in Lucerne, Ken is overcome with guilt and decides to call the chateau from the payphone at the café. He is put through to her room, where by this time she is fairly drunk, and before asking who is on the other end of the phone she blurts out, "Where is my second order of caviar?"

"Well, I guess you are ok and back to your old self," Ken says.

"You son of a bitch, who do you think you are?"

"Dawn, get a hold of yourself. I called because I was feeling guilty, and now I am convinced I did the right thing."

Ken hangs up; she starts to laugh.

"These little boys and these little games. They will see who the boss is."

She calls the concierge and tells them to have a car

brought up and to have the second caviar and vodka put in a to-go basket for her. She tidies herself up a bit and puts the cash in the secure safe in the room but takes a stack of one hundred dollar bills with her. Once in the car she asks the driver to summon the concierge, who comes to the window of the car.

"Is everything all right, madam?"

"Yes, I would like to know where the best dance clubs are."

"Yes, of course."

The driver is given instructions. As the driver pulls away he asks Dawn if she is enjoying her day.

"Yes, and I hope it gets even better. I enjoy hunting," she says.

"'Hunting,' madam?"

"Oh, that is just my way of saying having a good time."

"Oh, yes, of course. I will have you to some very nice places shortly."

They pull up in front of a very lively place that has a long line waiting to get in. The driver asks Dawn to please wait a moment while he has a word with the doorman. That being Shamus's last job causes her to think about her brother for a moment. The driver returns to the car to find Dawn looking straight ahead with a blank stare.

"Ma'am, I got you right in; you do not have to wait in this line."

Dawn is fixed in her gaze and does not reply to the him.

"Ma'am, are you all right?"

"Uh, yes, I was just thinking about something; did you say something?"

"Yes, you can go right into the club and not have to deal with this line."

She takes a healthy sip of the vodka that is with her in the car and he opens her door. Dozens of people are staring at her and the car as they realize she is being brought right in. Everyone is wondering who she must be. A couple of young men take particular notice and one of them says to give him five minutes and he will have her dancing. He says it loud enough for Dawn to hear, and she says, "Why will it take you that long?"

She is whisked past the line, leaving the young man even more motivated to get in. As she is led into the club, Dawn tells the driver to go back and have the young man sent in now. He does what he is told and the young man is singled out from the line; he asks the driver if his buddy can come.

"Sorry, but the lady just asked for you."

The other young man tells his buddy to go in and that he will surely catch up as the line is moving fairly quickly. He tells his lucky friend to have some extra drinks ordered and ready for him. The driver leads the young man in. He is a very handsome, tall, lean, blonde, Nordic-looking fellow. He is very well dressed in a finely tailored suit with an open collar shirt. He is brought to her table and thanks Dawn for getting him in.

"Of course, by all means," she says. "Here are the rules. I don't want to know your name and don't ask me mine."

"Are you sure you want some company?" he says.

"Did you come here for a good time or to ask questions?" she says.

He doesn't say a word and sits down with a big smile.

"May I buy you a drink, pretty lady?"

"No, this evening will be on me. What would you like? I have been having vodka."

"That would be great; on the rocks, please."

They have a VIP table in a roped-off section where the drinks come fast and furious and go down easily. The place is very loud and the band uses well-timed laser lights to heighten the experience. Hundreds of people are dancing, many of which try to push by the VIP area to see who may be there. Soon Dawn and her prey are completely drunk and making quite a scene. He has his hand up her dress and she is licking his chest hairs. By this point his buddy is finally in the club and sees this.

"You lucky bastard," he shouts over to him.

"Hey, come join the party!" Dawn says.

Soon the friend is drunk as well and all three are dancing and she starts to kiss the friend as well.

"Hey, let's get out of here and head back to my place, boys."

There is no argument about that and they head out to the waiting limo, where she instructs the driver to take them to her hotel.

As the limo pulls up to the chateau they are greeted by the concierge, who asks if she had a nice time.

"Yes, but I think the best fun is yet ahead," Dawn replies.

The three make quite a scene as they head to the elevator. They are so drunk they are bumping into things but manage to get into the elevator and head to her suite. It doesn't take any imagination to know what happens the rest of the evening.

While this has been going on, Ken has checked into a seedy hotel in Lucerne, not far from the train station, where he was able to pay in cash and use an alias to check into his room. His room smells of mothballs and the furniture is stained, broken, and torn. The guilt of drugging Dawn is overwhelming Ken's every thought, so he decides to put another call in to her, just to see if she is all right. Ken goes back down to the lobby to use the pay phone and is connected to Dawn's suite. She picks up the phone and shouts, "I am waiting for the next bottle of vodka! Where is it?" He can hear the men laughing in the background. They add to the order, shouting, "Make it three bottles!"

Ken does not say a word and hangs up. He cannot believe his ears, but is not surprised. He returns to his room and thinks about his next steps.

Chapter 21:
THE WALLS CLOSE IN

The Messinas wake up late the morning after the festivities. Damiano has called the Bianca home and needs to speak urgently with Matt. There is a knock on Matt's door. It is Rose.

"Damiano needs to speak to you right away, dad."

"Thanks, honey, I will take it in here."

Matt makes sure Rose hung up and asks Damiano what is so urgent. He tells Matt that the phone bugs have picked up critical information.

"It seems as though Dawn's lover has been stealing from the company. Shamus was dropped at the airport and Dawn went to the airport hotel, where she was monitored until they lost track of her. She is on the run and they are not sure who the lover is just yet, but are sure he worked in one of the Manhattan locations in the finance area."

"Let me get this right: you think she and her lover conspired to get money out of the company and were successful in doing so and are now fugitives?"

"Yes, that is it exactly."

"Well, it's not going to be so hard to figure who the guy is when he doesn't show up for work. Damiano, get my CFO's office on the phone with me right away."

"Hold on, Matt, I will conference Brad right away."

Brad's secretary brings him in on the call to join Damiano and Matt.

"Matt, how are you enjoying Italy?"

"I am sorry, Brad, but we have no time for small talk. Shut all of our operating accounts down and block access to any corporate funds immediately. We have evidence that Dawn and a lover have been stealing from the company and are on the run."

"Matt, did I hear you correctly; is this some sort of prank? I know you are a kidder."

"No, Brad, I am as serious as a heart attack right now, and I need you to do this right now and wait for further instructions from me."

"Yes, of course, Matt."

"Brad, one other thing: this needs to be handled

in a way that does not panic our staff and customers. Announce this as a routine systems conversion and instruct the staff to apologize for any short-term inconvenience. Meet with Human Resources. I will need attendance records of all employees sorted by location and sent to me by the end of the day. Please have Human Resources coordinate this with your office and legal; we will need to track this very carefully. Anyone that had access to account funds, especially accounts payable, I want their attendance reports over the last year and I want that report by the end of today."

"Yes, of course, Matt, I am on it."

"Damiano, I want you to head down to Brad's office and get the legal staff together for a call at 4 p.m. your time. Also, I want to speak personally to anyone who was tracking Dawn; let's make that a separate conference call. Have them on the phone with me at 3 p.m. your time. Depending on who this guy is and where they are we may want to handle this in a certain manner."

"I understand, Matt, and I will head to the New York office right now."

"Thanks, my friend, be safe and take care; we will need all our energy for this situation."

Matt hangs up the phone and goes to the window,

where he stands with his left hand against the window frame and the other clenched into a fist behind his back. As he thinks about these developments, he begins to smile, and then he breaks out in an almost sinister laugh. Rose has been listening through the door the entire time but could not hear what was being said; she is able to hear the laughter, however, and walks away.

Matt knows that it will be easy to identify the cheating thief, and with his resources he will be able to track him and Dawn down. Matt has clearly developed a plan but will keep it to only himself and Damiano for the moment. Matt knows that it is imperative to keep this out of the public eye as much as possible. The business of risk management and insurance business is based on security and trust. To have his own wife and an employee involved in some sort of conspiracy would likely cause severe disruption to the company's reputation and possibly worst.

He walks over to his bedroom door, opens it, and calls for Rose. She comes right away, and Matt asks her to bring the boys up to his room for a family discussion. She quickly rounds up Luke and Mark and the four are alone in the bedroom, where Matt explains the situation. He tells them that none of this can be spoken about outside of the family.

"I want you all to think hard about any possible clues or patterns you may know of with Dawn that can help lead us to her and possibly her lover."

"You know, I always wondered about the so-called charity that she was involved with; you know the one she would have meetings with all the time but when we asked about it she would simply say it was a local food bank or something like that," Luke says. "Her line was that she did not want any recognition but to just support it in some way."

Matt comments that Luke may be on to something and asks them all to keep thinking about things such as this but to keep it private among the four of them.

"We and the business cannot risk getting this out in public; the implications to the business and our family could be catastrophic."

"That bitch, that damn bitch! Mark blurts out. "I am going to get her myself."

"Stop it, Mark. As bad as this is, we must keep our cool; there will be plenty of time for venting, but now is the time to remain calm and focused and let our most trusted people deal with this," Matt says. "Damiano is heading into Manhattan at this moment and I have a couple of calls set up this afternoon with key staff members from finance, legal, and Human Resources. I will brief you all after that, but for now I want you to go about your business and not raise the specter of anything being wrong. I will tell the Biancas that I have important business to conduct and will be working from my room later today. I seem to recall that Giovanni was offering to take you all sightseeing today."

"That is correct, and we will likely be gone for dinner as well," Rose says.

"Perfect, you three go do that and I will know much more later. Look, I don't want the three of you worrying more than you should; we just need to be as tight as ever and we will get through this. Frankly, I think I may turn this into a good thing."

"How could this possibly be a good thing?" Mark asks.

"This will be a great learning experience, and if handled correctly, we can turn it into a positive. We are Messinas; we always come out on top. I have prayed about this already and I suggest the three of you do the same. Praying calms the soul and allows you to think clearly. God doesn't always give you what you want but if you listen carefully he always shows you a way."

Later that day Matt connects with the New York office on two separate conference calls. The first is with the investigative team that had been tracking Dawn and monitoring hers and Shamus's phones. He is told that Shamus boarded a plane for West Palm Beach and that Dawn was last seen at the airport hotel; they had since lost track of her. Matt questions how they could have lost track of her if she was simply at a hotel. They don't offer any solid reason other than to say that things are not perfect in the world of tracking people, especially in crowded, high-traffic areas. Most of the useful information was from the wiretaps. From the information they recorded, they are able to tell who Dawn's lawyer is and that money had been extorted from the company.

Matt asks Damiano to have those on the call in the New York office leave the room for a moment. Damiano confirms they have been removed and cannot hear the conversation. Matt tells Damiano that Shamus must be with his aunt in West Palm Beach and that Dawn probably sent him down there since he loves his aunt so much and to further isolate him being a liability to her. He tells Damiano not to tell anyone about the aunt and to head down to Florida to learn what he can from Shamus without letting the aunt know he is there. Damiano is then instructed to have the investigative team come back into the room. Matt tells them that they are to continue to try to locate Dawn and to report any findings immediately to Damiano and no one else. He thanks them for their ongoing service.

It is now time for the second call to begin. Damiano calls the legal team into the room. Matt makes it very clear to them that he does not want any of this to get out into the public domain. Matt's personal and business lawyers are in the room, so he tells them to collaborate and make sure that the business steps that are being taken, along with the investigation, don't expose him or the company to any liability. Brad is also in the room, along with the executive staff, and tells the group that the only person who did not show up for work that has account access was a man named Ken Wilkenson and that his computer and files are being gone through at the moment. In fact, he has not called in at all today. Matt explains that he is not too familiar with Ken, other than knowing that he was hired after being highly recommended by another business owner that he knows. Damiano asks for the Ken's file and looks at his picture.

"Well, we can't jump to conclusions just yet, but from the looks of this guy he would attract a lot of ladies and we will have the investigators get on his trail right away."

Matt informs the legal team what to listen to on the tapes from the wiretaps and to not approach Dawn's attorney just yet. Matt is told that her criminal activity will void her prenuptial agreement and not to worry about that. He makes it very clear to everyone that his only concern is the reputation of his family and the company and that he has already thought of some plans as they relate to Dawn and her so-called lover. Matt instructs Brad to have their accountant's forensics people work around the clock to determine how much money may be gone. Matt tells the staff he has no plans to return to New York and will monitor the situation from Sicily. Being with his family in such a special place far away from this drama helps him think and plan what has to be done. He wants to control things his way with as little drama as possible and Damiano will see to it that Matt's directives are followed to the letter. The legal staff is dismissed and Damiano remains in the room. He tells Matt he will leave for West Palm Beach immediately.

Damiano arrives in West Palm Beach at 11:30 p.m. and checks into his hotel, which is in walking distance of Aunt Pauline's home. Early the next morning he takes his breakfast in his room before heading out to see if he can spot Shamus. Everyone in Palm Beach dresses in bright colors and preppy clothing, so he has brought those with him as to blend in. He hates these clothes, as he is anything but a preppy guy, but he does his duty and walks out with bright green pants and a white button-down shirt along with a golf hat and tortoise-rim sunglasses; he is truly part of the local fiber.

His walk takes him past the hotel's golf course and over to the extremely well-groomed, palm-tree-lined streets of Pauline's neighborhood. As he walks by her house he notices her walking the dog, and not sure whether or not she will recognize him, he decides to turn down a side street for a moment before going back to see how far from the home she is. She is far enough away where he is comfortable with knocking on the door and seeing if Shamus is there. As he walks up to the door he can see that Shamus is in the side yard, sitting in a lounge chair. Shamus does not recognize him at first, but gets up as Damiano is within feet of him.

"Damiano, what are you doing here? I am scared. I didn't do anything; are you here to hurt me?"

"Calm down, Shamus, I would never hurt you. You are not in trouble so long as you tell me where Dawn and her boyfriend are."

"I don't know, but the last time they were together they were at his uncle's home."

"What is his name?"

"It is Ken and the home is in Princeton."

"Do you remember the address or street?"

"I know it was on Library Way and it is on the corner of another street and looks like no other home. It has a red tile roof."

"Listen to me, Shamus. You are not in trouble so long as you give me good information and do not tell anyone I was here. Do you understand me?"

"Yes, Damiano, I do."

"Do you remember anything else? Were they going anywhere or did they have any plans?"

"I am not sure; it was so confusing. I just wanted to get here to my aunt's home."

"Here is my phone number; you call me if you remember anything else or if you hear from Ken or Dawn, but I must repeat to you that you are not in any trouble so long as you tell me the truth and cooperate and don't tell Ken or Dawn we are onto them. They used you, Shamus, and I hope you understand this."

"I do, but I am scared."

"It is going to be all right. Just call me if you have any more information; I must leave before your aunt returns from walking the dog. Remember Shamus, not a word of this to anyone!"

Damiano heads back to the hotel. This time he takes a longer route away from the neighborhood. He checks out of the hotel and is on the next plane to New Jersey, where he lands late that afternoon. He drives down to Princeton and has dinner. After it is dark he drives past the home that Shamus has described and notices all the lights are off. He repeats the drive-by a few times to size things up and parks the car a few blocks away. By now he has changed out of his Palm Beach attire and is in his favorite black turtleneck and jeans. He walks to the back of the house, and after surveying it a bit more feels assured that the home is unoccupied. He is able to get through the back porch door with ease and finds the backdoor unlocked. He thinks to himself that they must have left in a hurry. After rifling through some paper on the kitchen table he finds handwritten notes for a flight number to Geneva. Along with those papers he finds a note with the name "Thord," an international phone number, and an address for a chateau.

Damiano does an efficient and expedient sweep of the premises. Nothing else of value is found as it relates to finding them and he quickly returns to downtown Princeton, where he phones Matt with the information from a payphone. Matt remains stoic and calm as he is told the information and instructs Damiano to get on a plane to Geneva as quickly as possible. Damiano tells Matt that he kept a copy of Ken's employee file with his photo that will be helpful once he gets there. Before they adjourn the call Matt tells Damiano to hang up and redial with Brad on conference. Once on the conference call the information that was found in the Princeton home is shared with Brad so that he can have the details as to advance his investigation and thus assist Damiano upon his arrival in Geneva.

Back at the chateau, Dawn is creating quite a scene with her exploits. She receives a call from the concierge, asking what time will she be ready to depart. Dawn tells the concierge to call Thord and tell him she is not leaving today and that he will have to deal with that. A short while later the phone rings in Dawn's suite with provision from Thord's office giving her two more nights only and to make the necessary arrangements to leave. Also she is told that the car service has been restricted to her departure only and that any more charges to the room will be her personal responsibility. This infuriates Dawn, as she is particularly tired of her finances being controlled by men.

She is hungover and tired from an evening of drinking and having sex with the two men who are still in the room but feel that their welcome has now worn thin as her anger begins to escalate. Dawn stomps into the bathroom to take a shower and the men use this opportunity to get dressed and make their exit without saying good-bye. She shouts from the bathroom, "I hope you boys are ready for round two." When she comes out with nothing on and realizes they are gone she goes into a rage. She smashes an empty bottle of champagne against the wall and throws herself onto the bed where she cries for a while as the realization of her isolation and fugitive status begins to sink in even more.

In true Dawn fashion, she calls room service and orders two double Bloody Mary drinks and some caviar. When the service arrives the staff sees the mess in the room and asks if she needs any assistance. In a stern voice she orders them out of her room and tells them she will let them know when she needs anything. She quickly downs the two drinks and passes out from the exhaustion of the prior evening and the effects of the alcohol.

That evening Damiano's plane arrives in Geneva and he calls Matt immediately for instructions. Matt tells him that their research has revealed that Thord is a Swiss banker and that the chateau is owned by the bank and typically used for clients of the bank. Matt further explains that Swiss banks are renowned for their culture of privacy—all the way down to the maintenance staff. None of them will share any info. Matt instructs Damiano to do some surveillance of the place until the legal team can give more advice on the international banking laws. Notwithstanding, Matt has an alternative plan in mind that he wants to keep to himself for a bit longer as more information is provided to him.

Once Dawn wakes up from her nap she plans to hit the town again. She calls down for the car service and is reminded that it has been restricted for her departure. She orders them to just put it in on her bill. By this time Damiano has been stationed just outside the gates of the chateau, where he sees Dawn strut out and into the limo. He follows the car into town, where Dawn hits several of the hottest clubs. She is seen leaving the last one with a young man and returns back to the chateau. Damiano remains vigilant just outside the gate, which remains open. He notices that the limo driver is out of the car smoking a cigarette. Damiano approaches the driver and asks about the woman he was driving. The driver says he is not at liberty to say, to which Damiano replies, "Are you at liberty for five hundred dollars?"

"Of course, I always associate 'liberty' with Americans," the driver replies.

Damiano shakes his hand with the money rolled up in it and the driver is happy to share what he knows.

"Well, she has been an utter mess around here. The word is that her boyfriend dumped her and took off. Ever since he left just a couple of days ago she has been out on the town picking up men and drinking to total oblivion. She has been asked to leave the hotel and as I understand it will be kicked out tomorrow or the next day. She has been disrespectful to everyone around here and acts like she is entitled and owns the place."

Damiano pulls a picture of Ken out of his jacket pocket.

"Is this the man who left her, and do you know where he went?"

"Yes, that is him, and he was taken to the train station."

"Do you have any idea where he went?"

"No, but the next train arriving was heading to Lucerne, if that is of any help to you."

"I am going to give you another five hundred dollars to monitor the woman for me. Would that work for you?"

"Yes, my American friend, of course; she has been a thorn in our side and we are happy to get rid of her."

"All right, then, here is my number and I will be at

the Hotel de la Suisse if you have any trouble reaching me. I am Damiano in suite 211."

Damiano returns to the hotel and advises Matt of these developments.

"Great work, Damiano; I am going to send you a couple of support folks to help divide and conquer. I want you to stay and deliver a message to Dawn while the others track Ken down. I will have them to Geneva on my private jet. There are a couple of gents I know here in the village from Sicily that will be perfect for the job. Do you remember Charlie and Gabe, the two that helped with the situation we had here some years ago?"

"Yes, Matt, they are right for this task."

"You may also need to get from place to place quickly, so it will be great to have the plane at your disposal. I do not want any law enforcement notified and we need to keep this as tight as possible. Let her know that she will not be prosecuted if she returns the money and leads us to Ken along with meeting with you in a location yet to be determined for more instructions. She and Ken need to learn a lesson and we need to keep this out of the public in that country as well. I will tell you what the lesson plan is once we have the money and the two of them in our grasp."

"I am all over it, Matt; we will button this up quickly."

"I know you will, Damiano, but please be careful; I wouldn't put anything past Dawn."

"Understood, Matt, shall do. I am going to head back up to the chateau and make sure she gets the message. Have Gabe and Charlie check into the Hotel de la Suisse and I will give them the details."

"Perfect, expect them to arrive this evening."

Damiano drives back out to the chateau and parks outside the gate; he walks into the lobby, where he is greeted by the concierge. He shows a picture of Dawn along with a hundred dollar bill and says he would like an envelope delivered to her room.

"Yes, sir, of course; we will do that right away."

Inside the envelope the note reads: "You are to come to the Hotel de la Suisse promptly at 8 a.m. tomorrow. We know you and Ken have stolen from the company and you will not be prosecuted under certain terms and conditions that will be presented to you in person. Should you discuss this note with anyone or retain legal representation and not arrive at the appointed time I will void this one-time offer."

The letter is slid under Dawn's door and announced

with a few knocks. She and her latest conquest are active in the bed and ignore the knock, saying, "Go away," without realizing there is no one there, just the envelope, which is visible from the end of the bed. After the two lovers exhaust themselves from rounds of drunken, sloppy sex, she notices the note on the floor and dismisses it at first, as she thinks it's her bill or a note reminding her she has to leave. Her lover gets up to go to the bathroom so she decides to go over and pick it up. She reads the note while standing near the door; it falls from her hand and she falls back against the wall in shock. She slides down to the floor without words. Her lover returns to see her there.

"What is wrong?" he asks.

"You had better leave; things are about to get ugly."

"Ah ha! I knew you were married. Is he coming?"

"Just get out."

He grabs his clothes and gets dressed; without saying anything he steps over her as he walks out the door. She picks up the note to read it a second time, rolls it up into a ball, and walks out to the veranda after pouring herself a glass of champagne. In a firm tone she states to herself: "They are screwing with the wrong lady."

In Lucerne, Ken is in his hotel room trying to come up with a plan to fade away with no trail. At this same moment a plane with Gabriel and Charlie is landing in Geneva. They are two men that you would not want to run into in a dark alley. Both are over six feet tall and have packed on a lot of pasta pounds over the decades. Charlie has pock marks like the surface of the moon on his skin from past ravages of acne. Gabe has a shaved head and a scar from his right lip to the bottom of his ear lobe. Damiano meets up with the two men in the hotel bar and they are posted on all the details and given a picture of Ken. It is decided that the two men will stay for the morning to support Damiano if Dawn shows up and to see if they can get any info out of her as to Ken's location. The men order a bottle of fine Amarone and toast to their successful mission. Early the next morning they meet in Damiano's room, where Gabe and Charlie are instructed to intercept her as she comes into the lobby and escort her to the gardens in the back of the hotel and not to connect him with the note. Since Dawn has never seen these two men it will be more intimidating and will not connect Damiano with the delivery of the message, which is important if Dawn were to take this in a bad direction.

Dawn arrives right on time and sees the two men approach her. Gabe says to please follow them into the garden. She says she is not going anywhere with people that look like them.

"You need not worry, my dear; we will be in a public place the entire time so nothing is going to happen to you, and as we see it, you don't have too many choices."

She pulls out a cigarette and removes her sunglasses for a moment.

"Yes, you are as ugly as I thought; it wasn't just these sunglasses. By the way, scar face, light this cigarette. One of you two goons must have something you carry that starts fires, right?"

"Light your own damn cancer stick; we are not here to serve you."

Once in the garden she is asked if she is going to live up to the terms in the note. She says she wants to make a deal.

"If I were to lead you to Ken and the money, would I be able to keep my prenuptial agreement in place?"

"Listen, sweetie, we were told no deals; take it or leave it."

"Deliver that message first and then get back to me. I don't have time for you crazy goons. I am going back to my hotel."

" Well, it's your life, sister, but you are not in a position to push anyone around," Charlie says. "By the way, we have people watching your every move, so you will not be going anywhere without us knowing."

"Good, watch my ass as I walk away, as you will never have a piece as good as this one."

Damiano is watching from afar and sees her leave. The men meet at the hotel bar and they tell Damiano about her request.

"You guys did tell her no deal, correct?"

"Yes, and she is quite the arrogant bitch."

"Let me deal with her; you two head for the train station and Lucerne to track Ken down."

Once at the train station they show his picture to the ticket agents on duty and one remembers him.

"I did see that fellow, but we are not permitted to give out any passenger information."

Charlie slides two hundred dollars under the window.

"Yes, I do recall that he bought a ticket for Lucerne on train 1150," the agent says quickly. "It's hard to forget a man that good looking."

"Give us two tickets on the 1150."

Within an hour the men are on that train and showing Ken's picture to the staff on the train. The coach master says, "Yes, he looks familiar, and walks away.

Gabe hands the man another two hundred dollars and asks, "How familiar?"

"I know he got off the train in Lucerne."

"Was there anything else you recall about the man?"

Seeing his reluctance, Gabe hands him another two hundred dollars.

"Yes, you have jogged my memory, sir. He was the luckiest man on the train that day. He was invited by one of our special passengers to join her for drinks."

"What do you mean by special?"

He turns to walk away, thinking he can get more money.

"Hey, buddy, I am not an ATM; cough it up."

"Well, all right, then. He was with Alessia Eversoll, a famous Swiss actress. She was attracted to him and they made eye contact and thus she asked him to join her for a drink."

"I see; where did she get off the train?"

"They both got off in Lucerne. Her limo was loaded by our staff and she left without him. I did overhear that he was invited to a party at her home and that's all the information I have for you."

Gabe goes back to his seat and informs Charlie that he has enough information to progress; they have to find out where an actress of the name of Alessia Eversoll lives.

Once they arrive in Lucerne Charlie heads up to monitor the area around her home while Gabe checks hotels with his picture. Ken is hunkered down in his hotel, racking his brain with what he should do next. He can't get his mind off Alessia, but knows that connecting with her would create far too much risk of exposure; yet, he wants to stay in touch. He writes her a memo, thanking her for inviting him to sit with her in the train and tells her that he will be in touch with her soon. He takes the note to a florist down the street and purchases fine roses to send to her along with the note and requests that they deliver them to her residence that day.

Of course he is paying for everything in cash, knowing any use of credit will most certainly expedite his getting caught. He has resigned himself to the fact that the power of Messina's organization will bring him to justice. He keeps coming back to the same conclusion: that he should contact Matt's legal team and make a deal.

At this point, Dawn is plotting to turn in Ken and he is doing the same to her; Ken was not cut out for this sort of thing. The pressure is just crushing his ability to form rational thought. He heads back to his hotel and resigns himself to call Matt's CFO from a payphone on the street. The call is put right through to Brad, where he is shocked to be hearing from Ken, who goes on to explain that he wants to return the money if the company will not press any charges. Brad says he will have to run that by Matt, as he would be the ultimate decision maker. Brad does not let on the fact that they have already connected the money with Thord and his bank. Brad asks Ken where he could be reached with the reply. Ken says he is stupid for doing this, but not crazy enough to give up his location.

"Well, that's understandable, Ken, but since you have already admitted this, how did you do it?"

"Brad, all you need to do is look at the reinsurance treaties we had with offshore insurers. A few of those were shell companies, so the premiums were never applied to underwriting any of our larger commercial clients."

"You son of a bitch. So this was not only our money, but the funds of many clients—and right under my watch. Perhaps I am the stupid one, Ken, for trusting the audits. I will be lucky if I don't get fired now, or even worse now. Call me back in one hour. I will relay this information to Matt. Before I do that you must assure me that every penny is coming back and give me all the details of the transactions."

Ken shares the details with Brad as well as informs him of the five million Dawn was given at the bank. Brad falls back in his chair and nearly drops the phone.

"Call me back in exactly one hour, Ken."

Brad immediately summons the senior attorney, Myron, to his office and briefs him and then calls Matt.

Brad is worried about the exposure he has as the senior financial officer in the company and so begins by telling Matt that he is personally embarrassed that this plot could have been hatched and executed under his watch and that Myron has been briefed and is here with them. Myron explains the risks of keeping this quiet versus prosecution and public trial of Ken and Dawn.

"It's a good damn thing we did not go public on the stock exchange and have all those regulators floating about," Matt says. "It is a miracle that no claims have arisen that involve the false reinsurance Ken constructed."

Matt gives the directive to his executives to keep this quiet.

"Every penny must be returned, and furthermore, Ken must cooperate with replacement of the funds to the proper policies and accounts in force."

Matt orders that they must not deal with any more offshore companies or captive arrangements until further notice.

"Call me back after you have given these terms to Ken. Also, tell him that he will be living in the company-owned Manhattan apartment and sequestered by our security people until the company confirms his repayment is made whole. As for Dawn, I will call Damiano, and not only will she have to return every penny she will have to void the prenuptial agreement. Call me back after speaking with Ken."

Ken calls at the appointed time and agrees to the terms but says he is not giving up his location for forty-eight hours as he needs time to take care of something. He is told that will not be acceptable, but that he will be allowed the time if he gives his location and consents to being guarded by men that have already been tracking him. He agrees to this and Gabe and Charlie are sent to his hotel within hours.

Ken wants this extra time to be able to visit Alessia.

The next morning he sends her another bouquet of flowers with a note to meet him at a certain restaurant that evening. He arrives there early while Charlie and Gabe position themselves at a table not far away to watch his every move. Her Rolls Royce pulls up right at the appointed time. She is recognized by people on the sidewalk passing by but is rushed in by the doorman. She is delighted to see Ken and he rises to pull out her chair.

"So, you managed to get away from the clutches of your clients, I see?"

"Yes, I did, and I am so glad you were able to come. They are enjoying the evening, and as the evening progresses Ken explains that he has to leave for New York and will be there on a special project."

He vows to return as soon as he is done with work if she would like to see him again. She is sad that he has to leave, but delighted that he will be back.

"Do you have any idea how long you will be gone?"

"I don't think it will be more than a month, and I will stay in touch, if that is all right."

"But of course; please do so! I am happy to drive you back if you like; my car is here."

"Would you please excuse me; I will be right back."

He goes to the men's room to think over his next step. Gabe notices this and tells Charlie he will be right back and joins Ken in the men's room.

"She wants to drive me home."

"No way, buddy, make something up; you are not getting out of our sight. We will be watching you, my friend."

Ken returns to the table and asks if she would like something for dessert. Detecting that something is bothering him, she asks if something is wrong.

"Yes, there is, Alessia. I am overwhelmed by your beautiful looks and personality, and sad that I have to leave."

She offers to drive him again.

"No, but thank you. I am only a short walk away and have to meet clients for drinks to plan a bit before they leave for New York."

"Ken, I know there is something special about you, and I want you to stay in touch with me."

He assures her that he will and walks her to her car and waves as she drives off. As soon as the car is out of sight Charlie and Gabe are out the door and right behind him to depart in a cab together.

"You are really something, buddy; you operate quickly, don't you."

He doesn't reply and turns his head away from Gabe and Charlie.

Chapter 22: The Solution

Back at the home of the Biancas, Matt is pacing back and forth in his room, thinking about the balance of prosecuting them versus handling this as an internal company matter. He understands the risks of both but keeps coming back to the same conclusion: prosecution would most assuredly cause too much negative press and loss of trust in the marketplace, which in the insurance business is mission critical. At this point the children have returned to the Biancas from their day out and come to Matt's room to check on him. Matt tells them exactly what is taking place and they can immediately see the stress he is under. Rose has an idea.

"Dad, you know you have asked us to pray about this? I have an idea. We should all take a walk down to the church together and pray together."

"I am very proud of you all; yes, let's do that."

The four of them walk down to the church, and on

the way they pass Sophia, who was out running an errand. She inquires where they are going and Rose says they are going to church to say a few prayers. Matt extends his hand to Sophia and asks if she would like to join them. She declines, as this seems like a family time. Matt explains that prayer is never restricted to family and can be shared with all people who care about God.

"We are going to pray silently and welcome you."

Sophia is grateful and asks for a moment to run home to get Salvatore. They all meet at the church steps a short while later. The church's doors are always open and they enter through the main door, where they all bless themselves with holy water that is in a marble font just inside the vestibule. As they walk up the center aisle, all six genuflect and turn to a side altar where they each light a candle. When it comes to Matt's turn, Sophia takes his hand and assists him in the process as the rest look on. Rose turns to look at her brothers as they all watch the obviously gentle interaction between Sophia and Matt. The presence of God is felt by everyone, and in that moment all of them feel like one family, connected in a most beautiful and spiritual way. There can be no doubt now that Matt and Sophia have been brought together in a most divine way.

After Matt's candle is lit, he takes her right hand in his and places his left hand on top of hers in a gesture of thanks and comfort. They all then return to a single pew, where all six kneel and lose themselves in silent prayer. Sophia cannot take her eyes off Matt. He senses her looking and turns to her and smiles. He silently thanks God for bringing his mind to peace and for his new friend and her son. His prayers bring him to a peaceful state. In his prayers he is reminded that love is a much more powerful emotion than hate. Revenge and hate are emotions that only wear down your soul. It is now that he is sure that prosecution is the wrong path, and although he would love to have revenge on Dawn, that would only cause more trouble. Matt's prayers have also given him an idea that Dawn should be taught a lesson about what it is to really care about someone and to see what people who really love each other can do. He knows it is over between them, but uses God's words to bring this horrible event to a better path.

They all conclude their prayers and walk out of the church together. Before the group parts, Matt tells Sophia that he has some business to take care of in New York. Rose will be staying behind, and the boys are welcome to stay and enjoy Sicily for a while.

"Sophia, please don't forget about the trip I promised you, Salvatore, and Father Passante. I will be in touch with you very soon to make the arrangements for that."

She takes Matt's hand and kisses him on the cheek and they say their good-byes and walk back to their respective homes. As they walk back to the Biancas' they stop at the fountain in the town square for a moment. Luke says that he and Mark should return with him.

"No, please stay here," Matt says. "As you can see, there is much to be learned in this place. Damiano and I have everything under control and I will feel better if the three of you are together right now, and Rose must stay her to finish her studies." The four circle in a hug and then return to the Biancas' home.

Once at the house, Matt calls Damiano to tell him of his plan for Dawn. Damiano is to make it clear to Dawn that Ken has confessed and they know where all the money is located. Furthermore, she should know that Ken is coming to New York under their control until all the money is back in the proper accounts.

"Not only will Dawn have to return the money and agree to nullify the prenuptial, she will have to work as an employee in the Messina foundation in Haiti for one year. She will be assigned to work to help house and feed the needy there, as well as learn to care for the sick. She will receive a stipend of one thousand dollars each week for one year and will report to the director there. If she is not agreeable to these terms I will pursue prosecution."

Within hours Damiano delivers this news to Dawn, and after becoming vitriolic at this offer she realizes that with Ken having given up so much information this is her only option and agrees to the terms. Matt is informed of this and summons his jet back to Sicily to return him to New York. Gabe, Charlie, and Damiano remain with Ken and take him to meet with Thord to arrange for return of the funds.

Matt is greeted at the airport by Brad, who brings Matt directly to the office. Matt tells him to not worry about his job and that he knows he had nothing to do with this but does make it clear that better internal audit procedures must be put into place. Brad has been with Matt from the very earliest days of the company and he kept those dedicated employees with him. Matt knows there is better talent in the market but values faithful dedication over most things as long as someone can do their job well.

Once in the office a call is set up with Damiano and Ken. Brad lays out the terms of the money transfer to the men as well as the details of Ken's transportation to the United States and his living arrangements in New York. Matt is also informed that Dawn has racked up twenty-three thousand dollars in uncovered expenses at the chateau. Matt is outraged.

"Fine, let's pay it, but I will reduce her stipend to five hundred per week for the year. One thousand was too mcuh anyway, and she can live just fine on five hundred each week in Haiti. In fact, she can stay in hospital housing for free as a volunteer."

Matt does not see this as vengeful but rather a way to potentially reform a deceitful evil person.

Upon her arrival in Haiti, Dawn is overwhelmed. She has never seen this type of poverty along with the struggles of people who have had so many troubles in their life. Matt and his first wife had started the foundation after going there themselves as volunteers and were touched by the families. The spirit and determination of the people there deeply affected the Messinas. The struggle for people to keep their families together was something that they could just not bear to sit by and watch while doing nothing. All of the Messinas were actively involved and each has spent time in Haiti working as volunteers. Each one of the Messinas learned many life lessons in Haiti, and as a result, never take anything for granted. The Messina creed is that everyone is here on earth to help and serve each other. They know that evil can never prevail over those who love and practice daily living with a servant's heart.

Dawn is escorted personally by Damiano to Haiti, where she is introduced to the hospital staff and the foundation director, after which she is shown to her room. She will share a very Spartan space no larger than a walk-in closet with a bunk bed and a small window. A staff bathroom is down the hall and is shared by many people. There is no air conditioning and she must share the room with a woman that is there from France as a volunteer. Her name is Lisette, and she is trained as a nurse. Lisette is a gentle and kind soul and immediately saw the pain and terror in Dawn's eyes. Dawn has been instructed by Damiano that she is not to say anything other than that she is there as a volunteer and did not want any special treatment just because she was a Messina. Matt's own family lived with the other volunteers when they were there. He felt strongly that this was part of learning what was truly needed and not gleaned from some report from the foundation management.

Dawn calls Shamus to tell him she has decided to volunteer in Haiti and will be there for at least a year to help the people there. Shamus is very confused by this but is happy that his sister is safe. She reminds him that the Messinas have the foundation there and promises him she will call every week to check on him. She explains that Matt and she are splitting up and there will be no trouble. After her time there is concluded she tells him that she would like to come to Palm Beach to live with Aunt Pauline and him. Shamus relays this to Aunt Pauline and she is overjoyed with the news, as she is getting older and would love the company. Pauline is not told of the reason for the divorce other than the marriage just did not work out.

Over the first few weeks, Dawn is taught the basics of caring for patients in the children's wing. Lisette takes Dawn under her personal watch and the two become very close. Lisette's husband died and her children had grown up and moved away so she had come to a decision to make her life more relevant by serving people in urgent need. Each night after a long, hard day, Dawn would cry herself to sleep while Lisette held her hand. It is obvious to Lisette that Dawn is not there of her own volition. Lisette does not prod Dawn for information, but rather supports her as a friend without the need to know more. Without Lisette, Dawn would likely go crazy and try to leave the island.

Dawn's jewelry tells a story of a woman who had come from great privilege. It was out of place in the hospital and many people would talk about it. Lisette had told her to hide it somewhere, but Dawn refused, as there was no place she trusted, even the local bank. After a long week at the hospital Dawn and Lisette decide to walk into town for a change of scenery. They have a nice dinner, and on the way back they are accosted by two young men carrying baseball bats. All they want is Dawn's jewelry. The ladies try to run but their efforts are futile and they are pushed to the ground. The men say they will not be hurt if Dawn hands over the jewelry, and so she does, but throws it down the street and says, "Go get it yourself, you bastards." They slap her in the face and the men quickly grab the goods and run off.

Lisette takes the crying Dawn in her arms and consoles her by telling her not to worry; the Messinas have so much power, they would find a way to get it back. Dawn is now compelled to tell Lisette of the events that led up to her being there. Dawn paints a false picture of what really happened. She describes Matt and his family as controlling, vindictive people who treated her poorly and says that she felt as though she was trapped inside the walls of an emotional prison. Dawn goes on to say that she was accused of stealing money from the family's business and that she was sent here as sort of a prison sentence. Lisette is a wise lady who listens intently but takes Dawn's story with a grain of salt; she does not let Dawn know of her doubt. The Messinas were known as kind, generous people, and many of Lisette's friends had worked alongside one of the Messinas as volunteers. Notwithstanding her disbelief of Dawn's diatribe of slander, Lisette recognizes that Dawn is quite alone and in need of a caring person. She helps Dawn to her feet and the two women walk back down the rocky dirt road to the hospital.

With each day the two women bond closer and closer as Lisette teaches Dawn how to care for the sick children. It seems as though Dawn's cold heart is starting to thaw as she develops relationships with coworkers, patients, and their families. There is one child in particular that Dawn gets very close to. Her name is Mahalia, which in Hebrew means "tender one." She had been made an orphan at a young age when her mother died of breast cancer and her father abandoned her after she became ill. It became apparent that the hospital was not the right facility for the child and that she would likely die if kept there. Lisette sits Dawn down one evening and suggests that, while Dawn is not in the good graces of the Messina family, they should know of this child. The foundation was known to be able to focus on special cases. Dawn knew she was right and called Damiano to tell him of this. Damiano told her that he would give the information to Matt.

Matt instructs Damiano to contact the medical director to get the details. The report at the hospital is in fact bleak, and states that if Mahalia is not taken to the right facility she will likely die within months. The foundation officials meet with the staff at the hospital and coordinate the arrangements for treatment in New York City. The day comes when Mahalia has to leave and she refuses to go without Dawn, so she contacts Damiano and pleads with him to allow her to come to New York with Mahalia.

"This beautiful young soul is only seven years old and has no one to support her. She sees me as a mother figure," Dawn says.

Dawn is given the approval to come to New York during the time of the treatment and is given use of one of the Messina company apartments while there.

On the way into the apartment building after a long day of caring for Mahalia, Dawn runs into Ken, who is being escorted by two men. Ken has been under constant watch during the process of returning the funds. She stops dead in her tracks and the two exchange looks like they had seen a ghost. He is rushed into the elevator as she stands locked in place, trying to catch her breath and process what just happened. She knew he was in New York for the return of the funds but she never expected to see him. Dawn has been left unescorted, as the Messinas felt that her concern for Mahalia was genuine and that her time in Haiti was drawing to a close soon anyway. They had her share of the money back and were no longer too concerned about her whereabouts. Nonetheless, they check on her frequently.

Mahalia was going through very difficult treatments and this was also hard for her. Damiano's people had come by a few times to check on Dawn while she was in the hospital and had reported that she was with Mahalia every second that she could be.

It does not take Dawn very long to start to sink her flirting claws into a handsome staff oncologist at the hospital, Dr. Jack Levinson. She saw him the first day Mahalia was admitted to the hospital, and since that moment she had her sights on him. He was a few years younger than Dawn and had striking good looks. His chiseled facial structure, dark wavy hair, and tall stature made him an imposing figure. His eyes were shades of bright green and blue, like the colors you would see in the Caribbean. He was attracted to Dawn not only because of her good looks but because of the way she cared for their mutual patient. This poor man had no idea what a pariah she was. All he knew was that she was the former wife of a very wealthy man and they were known to be philanthropists, so he assumed she had a servant's heart. He and Dawn grew closer with each day and started to date. The discussion each night was their mission to save Mahalia. Over the period of several weeks, Mahalia slowly got better and was about to be released when Jack extended an offer for Dawn and Mahalia to stay with him at his Upper East Side apartment. In front of his entire staff, he dropped to one knee and opened up a small box with a beautiful diamond ring and said, "Even better: would you marry me and let's adopt Mahalia?" Dawn is breathless and the staff is overwhelmed with emotion. She immediately accepts. As she and Jack hug and kiss over and over again, she cannot help but think about how she will be back in her old neighborhood. One can only use their imagination to see how this will play out.

During this time Ken is diligently working with Matt's executive team while under what he terms as house arrest. All the money is recovered and realigned to the proper accounts within eight weeks. Ken is released with words of caution to never speak of this again and he does not hesitate to return to see Alessia.

Once there the two bond quickly and are seen traveling all over Europe, where they are photographed by paparazzi in Paris, Monte Carlo, and Rome. All speculate they will be married within months.

Chapter 23:
BLIND VISION

Dawn's engagement is plastered all over the society pages of the New York newspapers and Ken's antics are not only in the papers but all the major people watching media publications. Damiano reads the latest news to Matt, who requests that he not hear about it anymore and asks Damiano to send a special invitation to Sophia. He has Alfredo make a heart-shaped pizza and has it delivered to her home with a note inside which says: "I have a beautiful blind vision of you in my heart and cannot wait to hold your hand again soon."

"Damiano, call the airport and have the crew ready to take me back to Palermo." Matt is ready for the next chapter in his life and wants to share it with the most beautiful, deserving soul he knows.